AF789676

PHYSICS RESEARCH AND TECHNOLOGY

NUCLEAR MAGNETIC RESONANCE (NMR)

THEORY, APPLICATIONS AND TECHNOLOGY

PHYSICS RESEARCH AND TECHNOLOGY

Additional books in this series can be found on Nova's website
under the Series tab.

Additional e-books in this series can be found on Nova's website
under the e-book tab.

PHYSICS RESEARCH AND TECHNOLOGY

NUCLEAR MAGNETIC RESONANCE (NMR)

THEORY, APPLICATIONS AND TECHNOLOGY

DODDAPUNENI KRISHNA RAO
EDITOR

nova publishers
New York

Copyright © 2014 by Nova Science Publishers, Inc.

All rights reserved. No part of this book may be reproduced, stored in a retrieval system or transmitted in any form or by any means: electronic, electrostatic, magnetic, tape, mechanical photocopying, recording or otherwise without the written permission of the Publisher.

For permission to use material from this book please contact us:
Telephone 631-231-7269; Fax 631-231-8175
Web Site: http://www.novapublishers.com

NOTICE TO THE READER

The Publisher has taken reasonable care in the preparation of this book, but makes no expressed or implied warranty of any kind and assumes no responsibility for any errors or omissions. No liability is assumed for incidental or consequential damages in connection with or arising out of information contained in this book. The Publisher shall not be liable for any special, consequential, or exemplary damages resulting, in whole or in part, from the readers' use of, or reliance upon, this material. Any parts of this book based on government reports are so indicated and copyright is claimed for those parts to the extent applicable to compilations of such works.

Independent verification should be sought for any data, advice or recommendations contained in this book. In addition, no responsibility is assumed by the publisher for any injury and/or damage to persons or property arising from any methods, products, instructions, ideas or otherwise contained in this publication.

This publication is designed to provide accurate and authoritative information with regard to the subject matter covered herein. It is sold with the clear understanding that the Publisher is not engaged in rendering legal or any other professional services. If legal or any other expert assistance is required, the services of a competent person should be sought. FROM A DECLARATION OF PARTICIPANTS JOINTLY ADOPTED BY A COMMITTEE OF THE AMERICAN BAR ASSOCIATION AND A COMMITTEE OF PUBLISHERS.

Additional color graphics may be available in the e-book version of this book.

Library of Congress Cataloging-in-Publication Data

ISBN: 978-1-63117-768-2

Published by Nova Science Publishers, Inc. † New York

Contents

PREFACE

This book is mainly focused on basic concepts and different applications of NMR from the small molecules to biological macromolecules. This book was made with contributions from different authors all over the world who are working on different aspects of science with the common tool of NMR.

Chapter 1 explains different types of NMR methods used to analyze the structures of small molecules and proteins. The concepts of different methods and how they will decode to spectral information to the structure is also explained.

Chapter 2 explains the complete story of how one can solve the structure of the new drug molecule. We explained sequential processes to identify the structure of small molecules using NMR.

Chapter 3 reviewed how the off-pathway intermediate during protein folding can be identified using NMR, in this we have included optical (fluorescence, CD, and UV-vis) spectroscopy data as supportive information for this study.

Chapter 4 is all about the how the NMR can identify and quantify the pure-exchange phenomenon in the protein. Specifically, it talks about aromatic ring flip motions in proteins under sub-denaturing and temperature variance.

Chapter 5 is a review on NMR structural and dynamical aspects of the death domain super family proteins. It explains how the super family proteins fold into similar secondary structures, even though their homology is about 10-20%. It also explains how these secondary structures play a role in the process of apoptosis.

Although NMR has established well in the solution state, solid state application and theory is still infantile. Chapter 6 in this book explains the emerging methodology and its theory. This chapter explains average

Hamiltonian theory, Floquet theory, Floquet-Magnus expansion, and Fer expansion for spin dynamics in solid state NMR.

Chapter 7 explains the metabolic profile and quantification of metabolites in PGI cherry tomatoes using solid state NMR. This chapter helps the reader learn how to quantify the metabolites in the solid state form of food materials.

Chapter 8 explains the different forms of single crystal using T1 relaxation in NMR. It also explains and compares the relaxation mechanisms and structural properties of different pure single crystals.

Chapter 9 focuses on the usefulness of Proton Nuclear Magnetic Resonance (^{1}H NMR) spectroscopy in the study of edible oils and fats, and of food lipids in general, from both qualitative and quantitative points of view.

Dr. Doddapuneni Krishna Rao
Magnetic Resonance Spectroscopy and Metabolic Imaging Group
Biomedical Science Institutes (A*-Star)
Singapore
Tel: +65-82470542
E-mail: kri_uoh@yahoo.co.in

ACKNOWLEDGMENTS

There is no word to thank people who dedicated their lives to research. We salute them, without their intense work and inherent wisdom we could not indeed make this book. Krishna Rao would like to thank all his teachers and supervisors who taught and impressed him at all stages of his career, he wants to extend his thanks to family and friends, who are always with him in all stages of his life.

In: Nuclear Magnetic Resonance (NMR) ISBN: 978-1-63117-768-2
Editor: Doddapuneni Krishna Rao © 2014 Nova Science Publishers, Inc.

Chapter 1

THE ROLE OF NMR IN STRUCTURAL ELUCIDATION OF SMALL MOLECULES AND PROTEINS

***Nadeem Khan*[1,*] *and Doddapuneni Krishna Rao*[2]**
[1]Jai Fesearch Foundation, Gujrat, India
[2]Singapore Bioimaging Consertium, A*STAR, Singapore

ABSTRACT

Since the early 1960's, Nuclear Magnetic Resonance spectroscopy (NMR) emerging as one of the most versatile techniques for structure elucidation of small organic molecules. With the evolution of improved experimental technology, practical strategies and novel approaches, NMR spectroscopy is paramounting the other spectroscopic techniques. This chapter covers the description of the strategies to analyze unknown compounds for the determination of structural connectivity, relative stereochemistry and verification of the proposed structure using 1D and 2D NMR approaches.

* E-mail: cht050@gmail.com; nadeem.khan@jrfonline.com.

1. INTRODUCTION

NMR involves the positioning of an atomic nuclei in an external magnetic field to generate two spin states with different energies. Low and High energy spin states nuclei absorb energy of radio frequency range to change its spin orientation as per applied magnetic field direction.

With the continuing development in processing techniques to abstract more and meaningful information from the experiments, NMR is one of the most powerful tool to elucidate the structure of small as well as large molecules.

Various types of experiments such as 1D-NMR (^{1}H DEPT, NOE, ^{13}C, ^{15}N, ^{19}F, ^{31}P, 29 Si etc.), 2D-NMR (COSY, DQFCOSY, TQF COSY, MQF COSY, HETCOR, HSQC, HMQC, HMBC, TOCSY, NOESY, EXSY etc.) and 3D-NMR (Homonuclear and Heteronuclear) were developed. The primary focus of this chapter is on interpretation of structures for small organic molecules by different NMR experimental techniques.

1D-Nuclear Magnetic Resonance spectroscopy:

a. *^{1}H NMR*: In proton NMR, spin transitions of different types of Hydrogen nuclei in different electronic environments were pragmatically observed. An interpretation of ^{1}H NMR data on the basis of chemical shifts is well understood and written. In Table 1, chemical shift values and coupling constant for various types of organic compounds are listed. All these typical values were observed on the scale of 0-15 ppm using Tetra Methyl Silane as a reference standard. A ^{1}H spectrum with a lesser number of scans can provide most of the spectral details. Using an optimum amount of the sample ($\geq$ 50 mm) dissolved in 0.6 ml of deuterated solvent in 5mm NMR tube gives good resolution. The volume of the deuterated solvent can be reduced to 0.4 ml to increase concentration, however one will need to spend more time on shimming to achieve better line shape. High concentration can introduce artifacts, which should be avoided. Example of 1H NMR is shown in Figure 1.

b. *^{13}C NMR:* Among the three isotopes of carbon, ^{12}C accounts for 98.9% and due to the absence of a nuclear magnetic moment, it is not a suitable nuclei for NMR and that is why it is considered as an NMR inactive nuclei. ^{13}C isotope with 1.1% abundance and nuclear magnetic moment ½, just like a proton, provides structural information for the organic compounds.

^{13}C NMR is now an extremely active area, which in the past considered as “least receptive Nuclei”. The ubiquitiousness of carbon in organic compounds and from the large chemical shift range relative to that of the proton, it gives hope for the spectral resolution of carbon, which also differs faintly in their chemical properties.
The properties of ^{13}C in comparison with other NMR active nuclei are shown in Table 2. Table 3 shows the correlation of the chemical shift for different organic functional groups. A chemical shift range of ^{13}C NMR is 0-220 ppm. an example of ^{13}C NMR is shown in Figure 2.

Table 1. Chemical Shifts and coupling Constants of ^{1}H Nuceli in different electronic Environment

Compound Group	Type of Hydrogen Nuclei	δ Value in ppm	J value
Alkane	$R-CH_3$	0.75-1.3	^{3}J=7-8 Hz
	$R-CH_2-R$	1.2-1.5	^{3}J=7-8 Hz
	R_3-CH	1.5-2.0	^{3}J=7-8 Hz
Alkenes	HC=C-H (*trans*)	4.5-6.5	^{3}J=11-18 Hz
	HC=C-H (*cis*)	4.5-6.5	^{3}J=6-15 Hz
	HC=C-C-H	1.6-2.5	^{4}J=0-3 Hz
	$HC=CH_2$(*gem*)	4.5-6.5	^{2}J=0-3 Hz
AlKynes	CH≡CH	1.7-2.7	--
	CH≡CCH	1.6-2.5	^{4}J=2-3 Hz
Aromatic	Ortho Coupling	6.0-8.0	^{3}J=7-10 Hz
	Meta coupling	6.0-8.0	^{4}J=2-3 Hz
	Para coupling	6.0-8.0	^{5}J=0-1 Hz
Aldehydes	R-C(=O)-H	9.0-10	--
	$^{\alpha}$CH-C(=O)-H	2-2.5	--
Ketones	-C(=O)-CH$^{\alpha}$	2-2.5	--
Carboxylic Acids	-C(=O)-OH	11-12	--
Esters	-CO-O-CH$^{\alpha}$	2-2.5	--
Amides	-CO-NH-CH	2.0-3.0	^{3}J=0-7 Hz
	CH-CO-N-	2.0-2.5	--
	-CO-NH	5.0-9.0	^{1}J=50 Hz
	-N-CH	2.0-3.0	--
Amines	-C-NH	0.5-5.0	^{1}J=50 Hz
	Ar-NH	3.0-5.0	^{1}J=50 Hz

Table 1. (Continued)

Compound Group	Type of Hydrogen Nuclei	δ Value in ppm	J value
Halogens	HC-I	2.0-4.0	--
	HC-Br	2.7-4.1	--
	HC-Cl	3.0-4.1	--
	HC-F	4.2-5.0	^{2}J=50 Hz
	HC-C-F	4.2-5.0	^{3}J=20 Hz

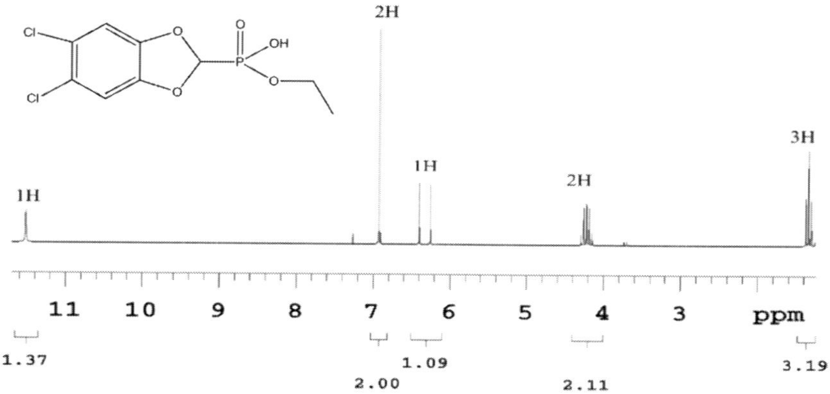

Figure 1. ^{1}H NMR.

Table 2. Characteristic properties of selected nuclei

Isotope of NMR active Nuclei	Natural abundance in %	Spin(I)	Magnetic moment (μ)*	Gyromagnetic ratio (y)**
^{1}H	99.984	1/2	2.7927	26.753
^{2}H	0.015	1	0.8574	4.107
^{11}B	81.170	3/2	2.6880	
^{13}C	1.108	1/2	0.7022	6.728
^{15}N	0.365	1/2	-0.2831	-2.710
^{17}O	0.037	5/2	-1.8930	-3.628
^{19}F	100	1/2	2.6273	25.179
^{29}Si	4.70	1/2	-0.5555	-5.319
^{31}P	100	1/2	1.1305	10.840

* $\mu = 5.05078 \times 10^{27} JT^{-1}$.
** $y = 10^{7}$ rad $T^{-1}sec^{-1}$.

Table 3. Chemical Shifts ^{13}C nuceli in different electronic Environment

Compound Group	Type of Hydrogen Nuclei	δ Value in ppm
Alkane	$R-CH_3$	15-30
	$R-CH_2-R$	22-45
	R_3-CH	30-58
Alkenes	$=CH_2$	115
	$=CH_2$ (Conjugated)	117
	=CHR	120-140
	=CHR(Conjugated)	130-140
	$C=C=CH_2$	75-90
	C=C=C	210-220
AlKynes	HC ≡ CH	65-70
	RC ≡ CH	85-90
Benzene	General range	115-160
Aromatic	Ortho NO_2 substituted	123.5
	Meta NO_2 substituted	129.4
	Para NO_2 substituted	134.3
	Ortho OCH_3 substituted	114.1
	Meta OCH_3 substituted	129.5
	Para OCH_3 substituted	120.8
	Ortho CH_3 substituted	129.3
	Meta CH_3 substituted	128.5
	Para CH_3 substituted	125.6
	Ortho Br substituted	131.9
	Meta Br substituted	130.2
	Para Br substituted	126.9
Aldehydes	RCHO	200
	ArCHO	190
Ketones	R_2CO	205-215
	5-Ring C=O	214
	6-Ring C=O	209
	ArCOR	190-200
Carboxylic Acids	RCOOH, ArCOOH	170-180
Esters	RCOOR, ArCOOR	165-172
Amides	$RCONH_2$, $ArCONH_2$	170
Nitriles	RCN	115-125

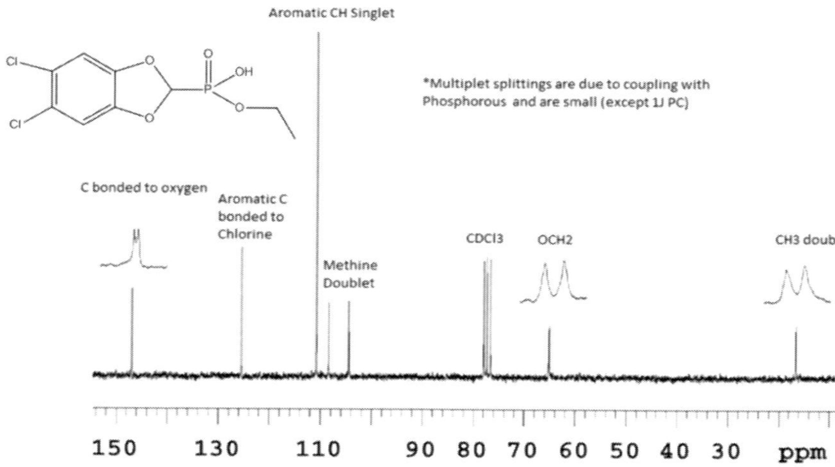

Figure 2.^{13}C NMR coupled with phosphorous.

c. *^{15}N NMR:* Nitrogen has two NMR active nuclei ^{15}N, which yield sharp lines, but insensitive and ^{14}N with a medium sensitivity usually broadened due to quadrupolar interactions sometime to the extent that they are unobservable even in a high resolution NMR spectrometer. Properties of ^{15}N and ^{14}N nuclei are given in Table 4. ^{15}N NMR chemical shifts for different functional groups are shown in the Table 5. The significance of nitrogen atoms in both inorganic and organic compounds and biological molecules makes nitrogen nuclear magnetic resonance a smart means of approach to study molecular structure and electron charge distribution. The negative value of the gyromagnetic ratio of ^{15}N may seriously impede an application of ^{1}H-^{15}N spin de-coupling for signal enhancement, which is a combined outcome of multiplet collapse and nuclear Overhauser effects, which will be discussed in the later part of the chapter.

d. *^{19}F NMR:* Natural compounds with fluorine are not abundant but in the chemical industry fluorine is the substituent of choice due to its strong ability towards hydrogen bonding (Vasle DINOIU, 2006). When a molecule has fluorine in the structure then it becomes a useful tool in structure elucidation. Due to resistivity towards degradation, fluorine acts as a probe for the determination of chemical residues and investigation of metabolic processes (Paul D Stanely, 2002). Magnetic properties of ^{19}F are listed in Table 6. A typical ^{19}F NMR is similar to that of ^{1}H NMR except wide range chemical shift from -450 to +440 ppm as shown in Table 7.

e. *^{31}P NMR:* Due to 100% isotopic abundance of ^{31}P and relatively high gyromagnetic ratio, ^{31}P NMR is a routine NMR technique, which helps in structure elucidation of phosphorous containing organic, inorganic compounds and metabolites. Usually chemical shift value of ^{31}P NMR ranges from -140 to 250 ppm as shown in Table 8 and the properties are shown in Table 9. The ^{31}P spectra is shown in Figure 3.

f. *^{17}O NMR:* With an abundance of 0.038%, ^{17}O is a naturally occurring isotope. It is a quadrupolar nucleus, giving too broad of a signal even with high resolution NMR spectrometer and also needs a high concentration to acquire the spectrum due to its low abundance. The chemical shift range for ^{17}O NMR is -40 to 1120 ppm as shown in Table 10 and its magnetic properties in Table 11.

Table 4. Properties of ^{15}N and ^{14}N Isotopes

Properties	^{15}N	^{14}N
Spin	1/2	1
Natural abundance	0.37%	99.63%
Chemical shift range	0-900 ppm	0-900 ppm
Frequency ratio	10.136767% or 10.1329111%	7.226317% (NH_3 7.223561%)
Reference compound	90% CH_3NO_2 or NH_3 Liquid	90% CH_3NO_2 in $CDCl_3$
Linewidth of NH_3	0.2 Hz	16 Hz
T1 of reference	5s	0.2 s
Receptivity relative to ^{1}H at natural abundance	3.85×10-6	1.01×10-3
Receptivity relative. to ^{1}H when enriched	1.04×10-3	1.01×10-3
Receptivity relative to ^{13}C at natural abundance	0.0219	5.74
Receptivity relative to ^{13}C when enriched	5.91	5.74
Electric Quadrupole moment (ex 10^{-24} cm^2)	0	1.54×10^{-2}

Table 5. Approximate nitrogen chemical shifts for organic compounds

Compound Group	Nitrogen Screening Constant relative to CH_3NO_2/NO_3^- in ppm
Nitroso compounds	-480/-540
Thionitrites	-350/-420
Nitrites	-180/-210
Nitrosamines (NO)	-150/-180
Azo Compounds	-130/-170
Nitroalkanes	0/-40
Nitroaromatic Compounds	20/0
Nitramines(NO_2)	40/10
Nitrates	60/10
Gem-Polynitroalkanes	60/10
Azoxy Compounds	60/40
Furoxans	70/-30
Nitrones	100/80
Thiocynates	110/90
Azines	120/-40
Nitramines(NR_2)	120/90
Azides (RNNN)	130/110
Azine N-Oxide	140/70
Nitriles	140/110
Nitrosamines (NR_2)	150/120
Fulminates (nitrile N-oxides)	180/150
Azinium ions	190/150
Cynamides	200/170
Isonitriles	220/170
Cynates	220/190
Thioamides	240/200
Nitrilium ions	250/220
Azoles (NR)	140/-10

Table 6. Properties of ^{19}F

Properties	^{19}F
Spin	1/2
Natural abundance	100%
Chemical shift range	-460 to +450 ppm
Frequency ratio	94.094011%
Reference compound	$CFCl_3$ =0 ppm
Receptivity relative to ^{1}H at natural abundance	0.83
Receptivity relative. to ^{1}H when enriched	0.83
Receptivity relative to ^{13}C at natural abundance	4716
Receptivity relative to ^{13}C when enriched	4716

Table 7. Chemical shift values of ^{19}F

Type of Compound	δ Value in ppm
-F-C=O	-70 to -20
-CF3-	40 to 80
-CF2-	80 to 140
-CF-	140 to 250
-ArF-	80 to 170
Xe-fluoride	250 to 440
FCl	-450
-RF-	-200 to -270

Table 8. Chemical Shifts ^{31}P nuceli in different electronic Environment

Compound Group	δ Value in ppm
$P(Alkyl)_3$	-62 to +63
$P(Alkyl)X_2$	186 to 285
$P(Me)H_2$	-163.5
P(Me)2H	-99
P(Me)2F	186
P(Me)2Cl	-96.5
P(Me)2Br	-90.5
$Ar_3P{=}CR_2$	5-25
$P(OR)_5$	-75 to -5
$O{=}P(OR)_3$	-20 to 0
$CP(=O)(OR)_2$	-5 to 70
$CP(=S)(OR)_2$	80 to 110
$CP(=O)(OH)_2$	-5 to 25

Table 9. Properties of ^{31}P

Properties	^{31}P
Spin	1/2
Natural abundance	100%
Chemical shift range	-140 to 250 ppm
Frequency ratio	40.480742%
Reference compound	85% H_3PO_4 in H_2O =0 ppm
Line width of Reference	1 Hz
T1 Reference	0.5s
Receptivity relative to ^1H at natural abundance	6.63×10^{-3}
Receptivity relative. to ^1H when enriched	6.63×10^{-3}
Receptivity relative to ^{13}C at natural abundance	37.7
Receptivity relative to ^{13}C when enriched	37.7
Nuclear magnetic moment	1.13160

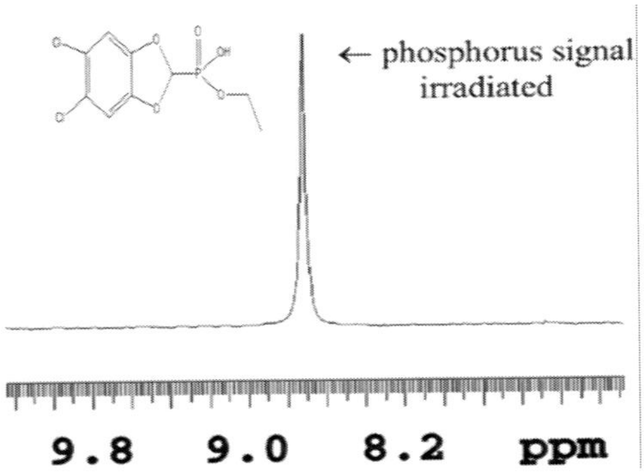

Figure 3.^{31}P signal.

Table 10. Chemical Shifts ^{17}O nuceli in different electronic Environment

Compound Group	δ Value in ppm
-OH	-50 to +70
-O-	-40 to +90
Acetales	10 to 100
-O-O-	170 to 280
Organic Sulphates	85 to 160
CO_3^{-2}	170 to 180
-COOH	225 to 250
Amide	265 to 320
Ketone	545 to 580
Aldehyde	545 to 625
Metal carbonyl	340 to 400
Nitro	410 to 440
Chromate	325 to 1120

Table 11. Properties of ^{17}O

Properties	^{17}O
Spin	5/2
Natural abundance	0.038%
Chemical shift range	-40 to +1120 ppm
Frequency ratio	13.556457%
Reference compound	D_2O

Table 11. (Continued)

Properties	17O
Line width of Reference	69 Hz
T1 Reference	0.02s
Receptivity relative to ^{1}H at natural abundance	1.1×10^{-5}
Receptivity relative. to ^{1}H when enriched	0.0291
Receptivity relative to ^{13}C at natural abundance	0.0650
Receptivity relative to ^{13}C when enriched	171
Nuclear magnetic moment	-1.8938

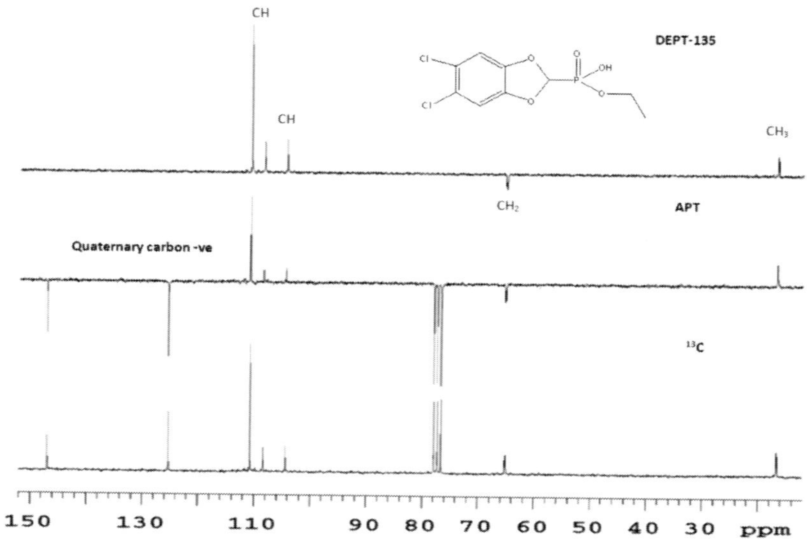

Figure 4. ^{13}C, DEPT and APT comparison.

2. Skeletal Connectivity

a. *Distortionless Enhancement by Polarization Transfer (DEPT) and Attached Proton Test (APT)*

DEPT is an advanced and improved version of insensitive nuclei enhancement with a polarization transfer (INEPT) experiment, which is based on relative population after selective inversion to observed insensitive nuclei with a larger intensity. Both APT and DEPT experiments are used to determine the number of protons attached to a given carbon atom. DEPT is a two channel experiment and employs a complex sequence of pulses in both ^{1}H and ^{13}C channel. The selection of pulses is purely depending on their flip angles and their delay times. During DEPT three experiments are recorded, where pulse has flip angle of 45°, 90° and 135° and the respective signals reveal the type of carbon atoms substituted. In DEPT 45, all signals are positive, in DEPT 90, only signals of CH will be observed and in DEPT 135, signals from CH_2 will be negative, while CH and CH_3 are positive. The APT experiment is one of the simplest multiplet detection techniques used in ^{13}C analysis. It is used to differentiate between proton attached carbons with quaternary carbon atoms. In APT methine and methyl signals are positive, while methylene and quaternary carbon signals are negative. APT is less sensitive than DEPT but reveals information on all types of carbons unlike DEPT, where signals for the quaternary carbon are missing. A comparison of ^{13}C, APT and DEPT is shown in Figure 4.

2D Nuclear Magnetic Resonance Spectroscopy:

b. COSY: It is a homonuclear 2D NMR correlation spectroscopy, which gives data plotted in a space defined by two frequency axes. It is used to identify spins coupled to each other such as the correlation of chemical shifts of two hydrogen nuclei located at two different carbons that are separated by a single bond through *J* coupling thus detect the chemical shift for hydrogen on two axis. COSY experiment is categorized into two types based on pulse angle used i.e., COSY-90 and COSY-45 using pulse angle 90° and 45°, respectively. COSY-45 is simpler and sensitive as it gives less pronounced diagonal peaks and provides cleaner spectrum. Another related COSY technique used are as follows:

Simple COSY: It involves simple pulse sequence in which, 90° pulse introduced in *x*-axisto one channel for example ^{1}H channel, to create achemical shift evolution. Thereafter, a second 90° pulse in *y*-axis is introduced to create an acquisition phase. In 1H-1H COSY pulse sequence holds variable relaxation delay time (t_1) and acquisition time (t_2) and the experiment repeated with different values of t_1 and t_2 such that the value of t_1 increases at regular intervals to generate a series of different FID data during t_2 phase. COSY technique provides three bond coupling data ($^{3}J_{H\text{-}H}$). Figure 5 shows , ^{1}H-^{1}H COSY of phosphonic acid ester where, *x*-axis (1D) is for ^{1}H channel and *y*-axis (1D) is for other ^{1}H channel and the central square with contours is 2D spectrum, which is a combination of two 1D spectra. We see a singlet and three multiplets in both proton spectra. In Central Square there is set of four signals along with a diagonal. When interpreting ^{1}H-^{1}H COSY spectrum the first thing is to draw a diagonal line across the spectrum from left and down to right top to identify the signals lying on it as shown in figure 5, where all the four signals i.e., one singlet and three multiplets observed lying on diagonal. Now we need to find out the coupling between protons. The doublet observed is due to proton-phosphorous coupling, so this coupling will not be active in ^{1}H-^{1}H correlation. You can see that –CH_2 and –CH_3 protons showing cross peaks or off diagonal peaks are connected by small square inside due to symmetry, which is a way of interpreting COSY spectra.

DQF COSY: It is the modified version of COSY that incorporates a distinctive pulse sequence in which, the first 90° pulse along *x*-axis introduced in ^{1}H channel, after another 90° pulse along *y*-axis with just the immediate third 90° pulse along *z*-axis is introduced to eradicate singlet peaks. Simple COSY and DQF COSY spectrum for ethyl acetate are shown in Figure 6a and 6b while phase sensitive diagrams of COSY and DQF COSY for serine are shown in Figure 6c and 6d. The difference in simple COSY and DQF COSY can be best understood by missing singlet for isolated methyl group in ethyl acetate.

TQF COSY:^{1}H-^{1}H TQF COSY uses additional typical sequence pulse in which, initially a 90° pulse on *x*-axis is introduced in proton channel followed by second, third and fourth 90° pulses on *y*-axis, *z* and z_1-axis, respectively to eliminate singlet and doublets.

MQF COSY: ^{1}H-^{1}H TQF COSY uses initially a 90° pulse on *x*-axis is introduced in proton channel followed by second 90° pulses on *y*-axis, immediately followed by multiple number of 90° pulses on Z_m-axis to eliminate triplets and any unwanted multiplets to make present data in simplest COSY spectrum.

Rules for predicting COSY connectivity:

1. The weak, medium and strong intensity levels of contour can be treated due to long range and short range coupling, respectively.

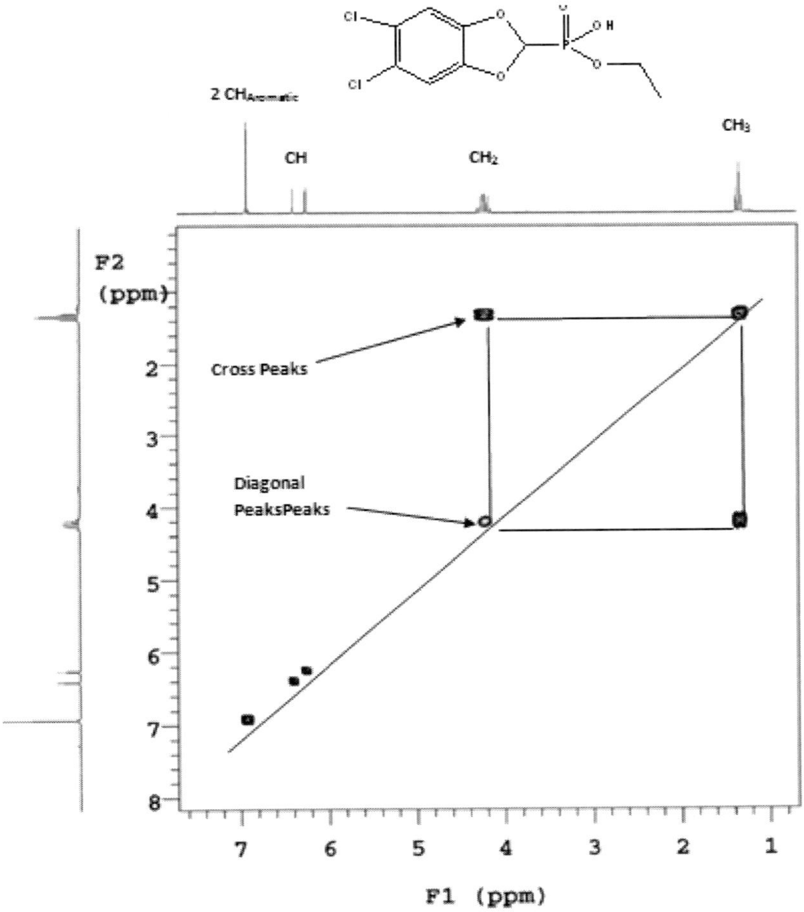

Figure 5. ^{1}H-^{1}H COSY spectra showing cross peaks.

2. If the intensity level is unknown then the J-coupling constants used to classify short range and long range couplings.

c. HETCOR or CH COSY: Heteronuclear 2D NMR correlations resonance experiment, usually between ^{1}H and ^{13}C resonances mediated by $J_{C\text{-}H}$. The experiment can be run for $^{1}J_{C\text{-}H}$ or for longer range couplings. It detects ^{1}H chemical sifts on F_1 axis and other heteronuclei chemical shifts on F_2 axis. When heteronuclei is ^{13}C then it has poor sensitivity and has been largely replaced by inverse detection experiments, which will be discussed later. Results of

HETCOR observed by locating cross peaks. If no cross peaks observed or detected, means no hydrogen present over carbon, one cross peak indicate the presence of one, two or three hydrogen nuclei, respectively. Two cross peaks reveals the presence of Hydrogen attached to distereotype carbon nuclei. Figure 7 shows the HETCOR spectrum obtained for Ethyl-2-Butenoate. The correlation between carbon and proton can be obtained from the spectrum and Contour "a" shows the correlation between the proton of $-OCH_2$ and carbon at 60 ppm and contour "b" indicates the correlation of proton from methyl group attached to the carbonyl in the acid part of the ester and not the methyl group attached to $-CH_2$ in the alcohol part of the ester. It was observed that carbonyl group from the ester has no correlation as it has no attached proton.

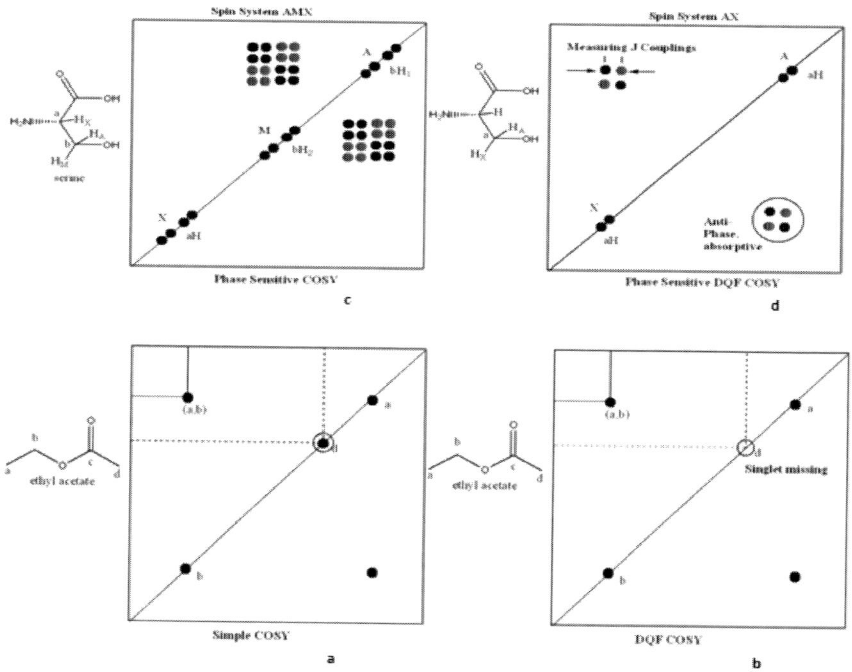

Figure 6. (a) Simple COSY of ethyl acetate (b) DQF COSY of ethyl acetate (c) Phase sensitive diagram of COSY showing AMX spin system for Serine (d) Phase sensitive diagram of DQF COSY showing AX spin system for serine.

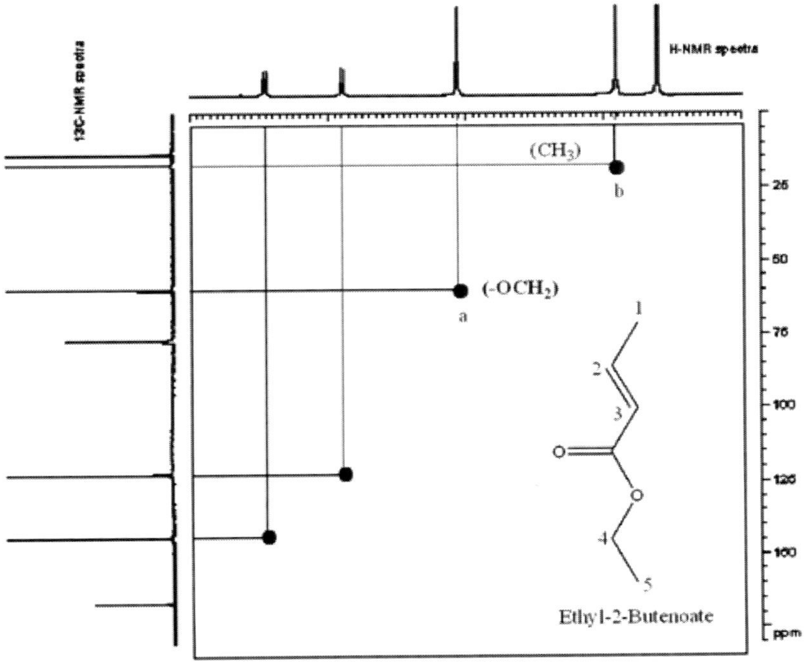

Figure 7. HETCOR spectrum of Ethyl-2-Butenoate.

d. HMQC and HSQC: 2D Heteronuclear multiple quantum correlation gives an idea about connectivity between ^{1}H and heteronuclei (mainly ^{13}C and ^{15}N). It is widely used and it is a proton based detection when compared with conventional HETCOR experiment. Carbon decoupling usually performed during ^{1}H acquisition and the corresponding satellite collapse to single resonance showing ^{1}H-^{1}H couplings. Heteronuclear single quantum correlation (HSQC) is another similar technique but better than HMQC in terms of resolution of ^{13}C peaks as shown in Figure 8.

e. HMBC: Heteronuclear multiple bond correlation is the modified version for HMQC suitable for long range ^{1}H-^{13}C coupling, it can be two bond, there bond and four bond as in conjugated system among which, three bond follows the karplus relationship i.e., for dihedral angles near 90°, the coupling is nearly zero.

HMBC produces cross peaks from ^{1}H and ^{13}C with coupling constant of 5-8 Hz. When experiment is design to suppress one bond correlation it become more attractive and the experiment is known as CIGAR. Interpretation of HMBC can be understood by Figure 9.

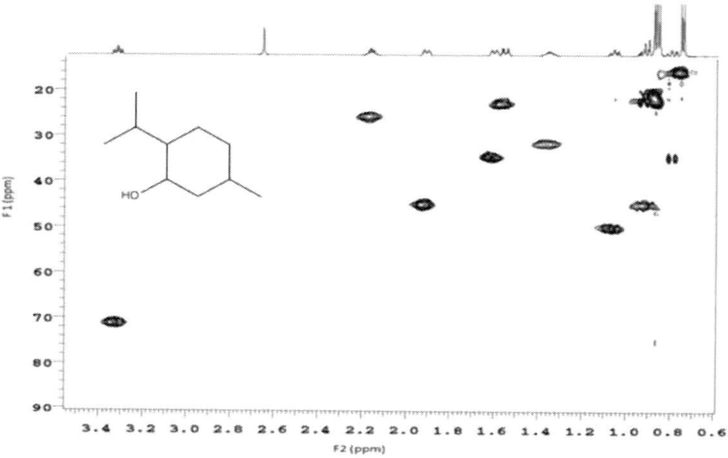

Figure 8. HMQC spectrum of Menthol.

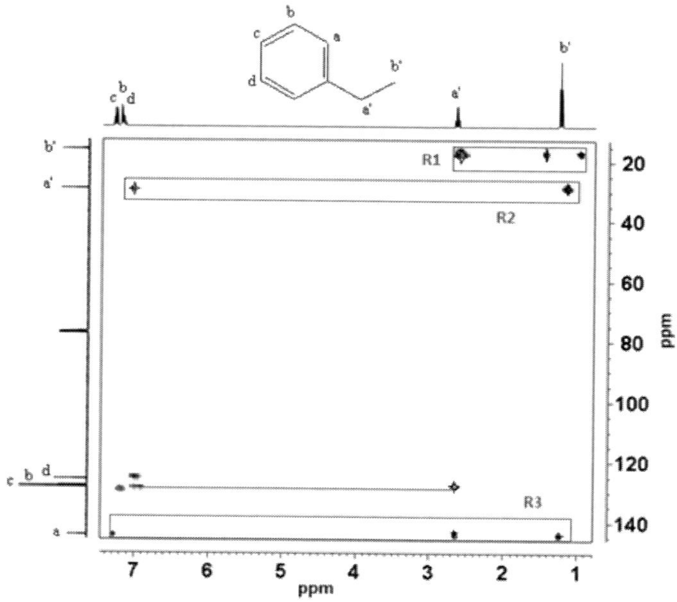

Figure 9. HMBC spectrum of Ethyl benzene.

In spectrum rectangle 1 (R1) parallel to x-axis, showing relationship between H_a and $C_{b'}$with two H attached on C_a. Rectangle R2 correlates $C_{a'}$ to Hydrogen attached to C_b. Similarly C_a correlated with hydrogen attached to $C_{a'}$, $C_{b'}$ (R3) and C_c; C_b with C_c Hydrogen, C_d and C_a: C_d with hydrogen of C_c and C_b.

f. TOCSY: Total Correlation spectroscopy as a homonuclear 2D experiment shows connectivity between all protons in a given spin system in germinal and vicinal as in COSY using transfer of magnetization successively over up to five or six bonds as long as the protons are coupled. The presence of heteroatom can disrupt TOCSY. Another name for TOCSY is HOHAHA (HOmonuclear Hartman Hahn). The TOCSY interpretation can be understood by Figure 10. It contains diagonals as well as cross peaks, the diagonals consist of 1D spectrum with single peak suppressed and the cross peaks indicates coupling between two multiplets i.e., CH_2 and CH_3 of ethyl benzene. All the desired signals corresponding to connectivity are in pure phase. Some artifacts also there but are rarely in phase with the desired signals.

g. EXSY: The 2D Exchange spectroscopy experiment is used to find out chemical or pure exchange between nuclei. EXSY interpretation can be understood from Figure 11 for N, N-dimethyl acetamide. In this molecule due to slow rotation around $CONH_2$ bond the two methyl groups exchanges with another rotationally however, actually shows different peaks in spectrum assigned as “a” and “b”, which is evident by the cross peaks in the spectrum.

h. NOESY: Nuclear Overhauser effect is a 2D spectroscopy method, which aim to identify the spins, which are connected through space. NOESY experiment is to check nOe in order to differentiate between two forms of carboxylate derivative. All peaks were assigned on the basis of ^{13}C APT, HMQC and HMBC spectra and on the basis of NOESY observed in –OH at 10.41 ppm and -CH (2.64-2.65 ppm), which confirms the form 2 of compound as shown in Figure 12.

3. Determining Relative Stereochemistry

By checking Nuclear Overhauser effect and coupling and decoupling experiments the relative stereochemistry of the molecule can be find out.

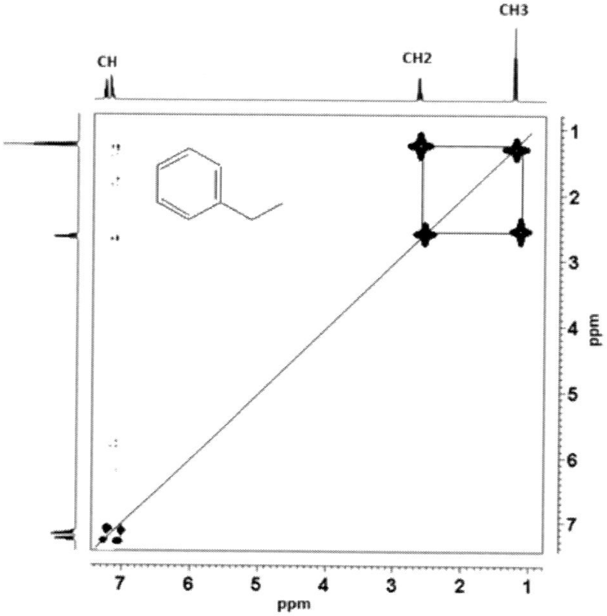

Figure 10. TOCSY spectrum of Ethyl Benzene.

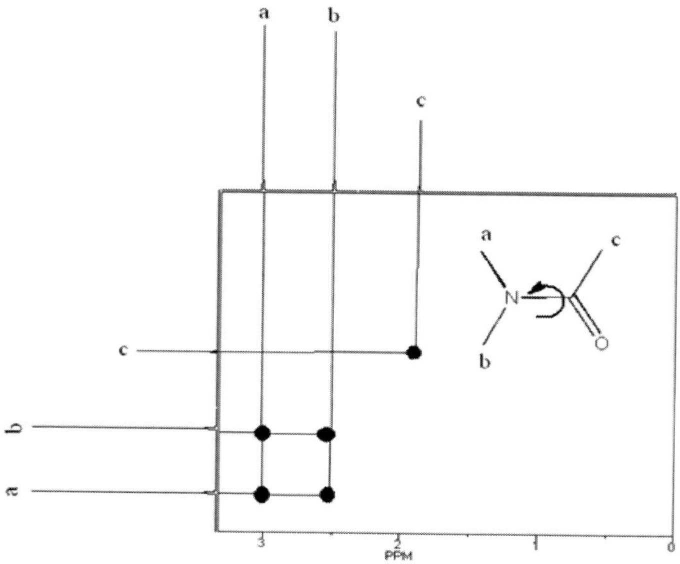

Figure 11. EXSY spectrum of N, N dimethyl acetamide.

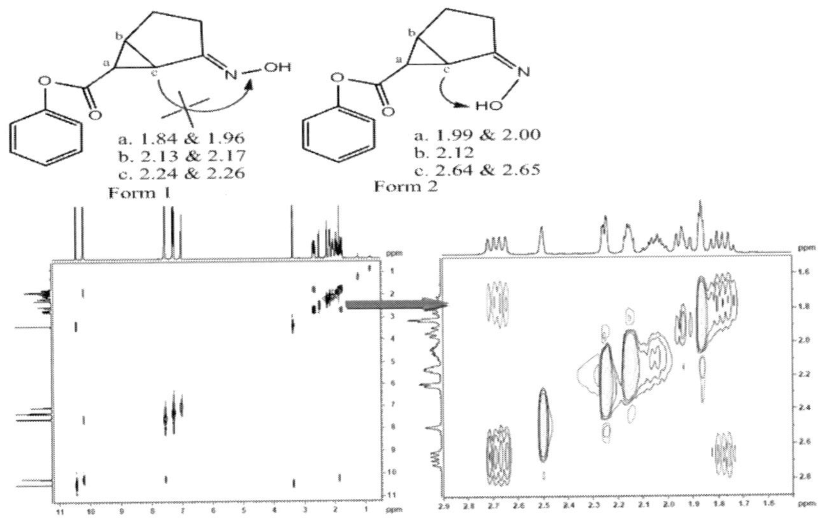

Figure 12. 2D NOESY spectra of carboxylate derivative for the confirmation of forms.

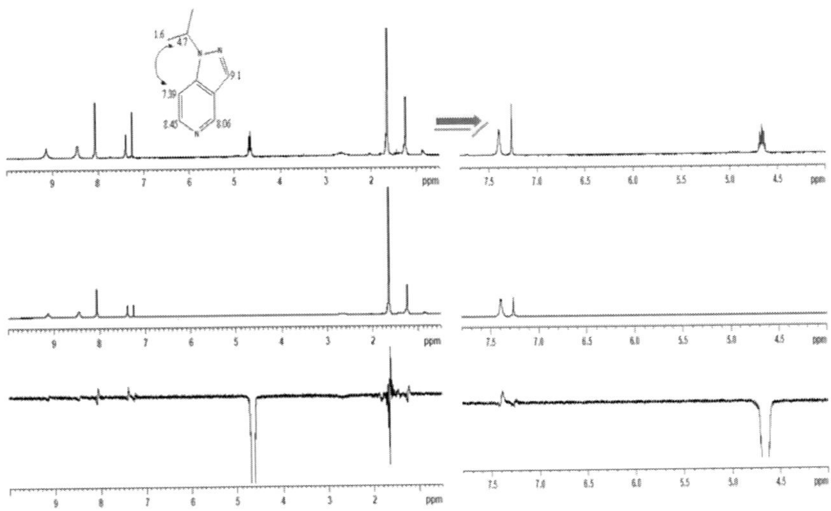

Figure 13. 1D NOE spectrum of Pyridine derivative.

1D NOE: Selective resonances were irradiated, which changes the intensity of nearby resonances and this effect is known as nuclear Overhauser effect and it is a powerful method of determining relative stereochemistry.

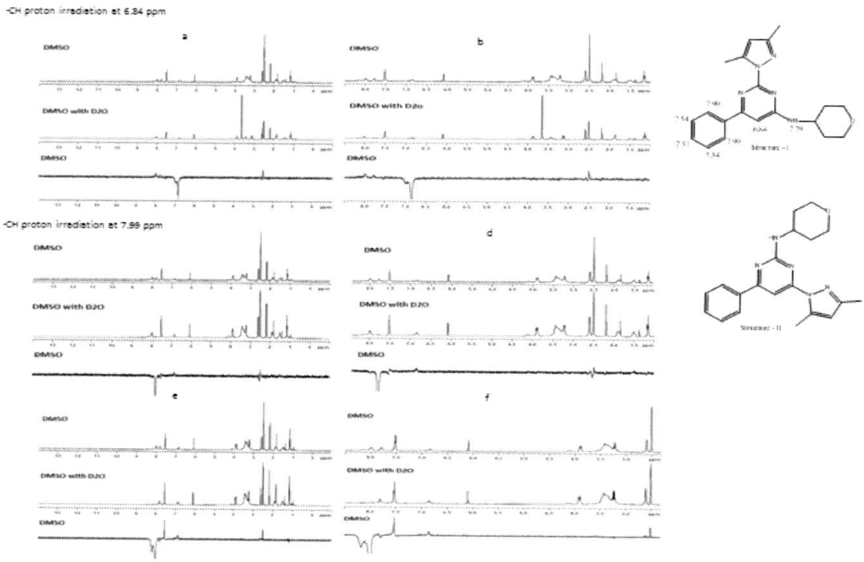

Figure 14. NOE spectra, a to f (to differentiate between Structure-I and Structure-II.

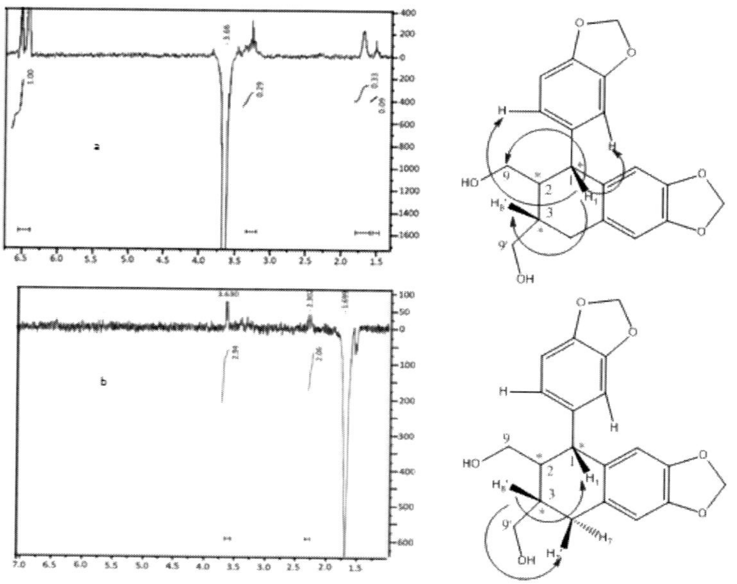

Figure 15. (a) NOE experiment upon irradiation H-1 proton at 3.66 ppm (b) NOE experiment upon irradiation H-8' proton at 1.699 ppm.

Interpretation of NOE spectrum can be understood from Figure 13 and NOE observed between CH (7.39) and CH (4.7). Another example is to locate the position of methyl group Figure 14a to 14f.

Upon irradiation of -CH proton at 6.84 ppm, enhancement in the resonance of -CH proton at 7.99 ppm observed and to confirm the proximity, -CH proton at 7.99 ppm irradiated and an enhancement observed at 6.84 ppm for -CH proton. Results obtained from NOE experiment indicate the probable structure for this sample is Structure-I.

4. Determination of Absolute Stereochemistry

The determination of absolute chemistry is also an important experiment in NMR to differentiate chiral isomers. Use of Mosher's Salt α-methoxy-α-trifluoromethyl-α-phenyl acetic acid (MTPA) has been most extensively used for derivatization of chiral alcohols and amines.

Diastereotypic protons will have different chemical shifts, but on the basis of chemical shifts two diastereomers can not be identified or confirmed and geometry may place certain protons in shielding or de-shielding portions of functional groups.

Figure 15a and 15b shows assessments of *trans*-stereo-configuration by homonuclear NOE experiments. Upon irradiation of H-1 at 3.66 ppm, enhancement in the signals at H-9 proton at 3.41 ppm and H-8' proton at 1.68 ppm observed but no enhancement observed on H-8 proton, thus confirming H-1 and H-8' are in close proximity, facing each other in space. To confirm the result of first NOE, H-8' proton irradiated at 1.69 ppm, an enhancement in signal of H-1 proton at 3.63 ppm observed along with effect on H-7 proton at 2.30 ppm. After combining results from NOE experiments, *trans*-configuration is confirmed.

Although the use of bond correlation spectra set up structural connectivity is likely to remain standard procedure, newer technology may soon improve the use of NOE and coupling constants for the determination of relative configuration.

New spectroscopic methods are continually emerging and with the help of availability of large empirical databases prediction of splitting patterns becoming easier and explanation of reliable automated structure are now well within reach.

5. Structural Analysis of Protein by NMR

Protein is made up of amino acids with a sequential connection through an amide bond. Proteins have a critical role in many biological processes. The relationship between protein sequence, protein 3D structure and its function is a challenge for biochemist, molecular biologist, and structural biologist.

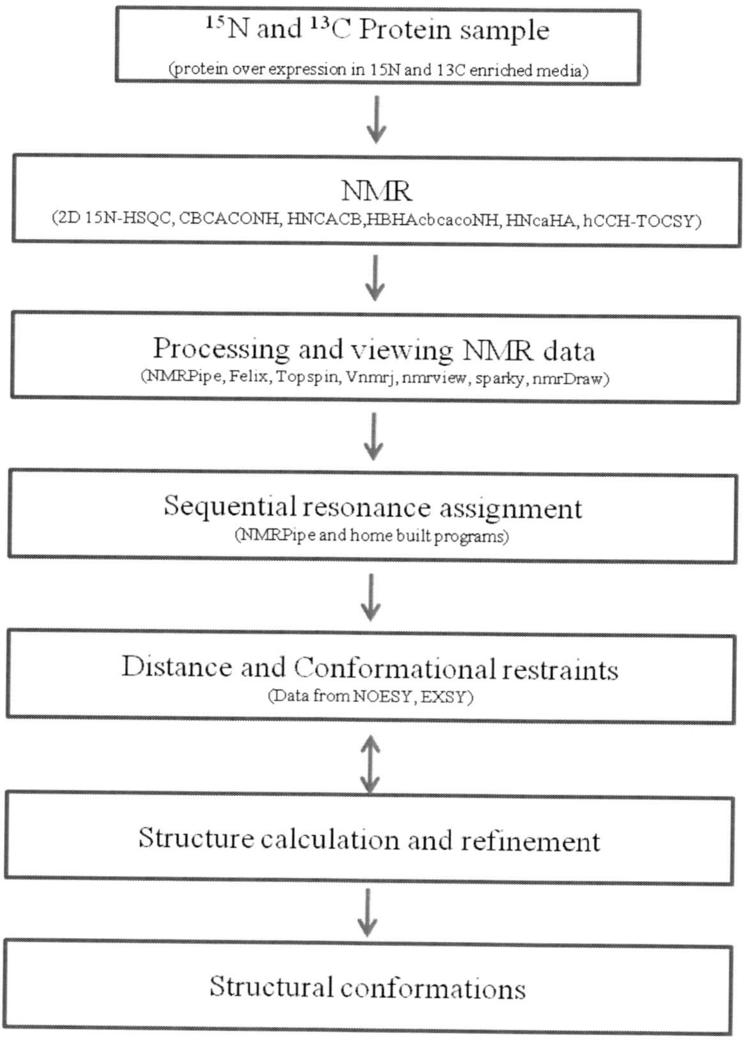

Figure 16. Schematic diagram for structural analysis of protein molecules by NMR.

The 3D structure of proteins is essential to understand the physical, chemical and biological properties of proteins (branden and Tooze 1991). Although the first atomic resolution structure of protein was done by X-ray crystallography (Drenth 1994), we cannot assume the same structure in solution state as it present in the body. In order to get more comparable structures to *in vivo*, it is essential to solve structure in solution state, which will be possible with NMR. The pioneers Abragam and Wuthrich (Abrgam 1961, Wuthrich 1986) were able make it possible to solve solution state structure of proteins. Although the NMR tool is used for protein structure, it is not a tool to produce direct image of protein, the process needs collection of enormous amount of data and analysis. NMR has its own disadvantages, like the nucleus should be magnetically active, abundance, and sensitivity.

To start with structural analysis of protein, we need to have isotopic (^{13}C, ^{15}N) enriched protein. So, as a first step one has to over express the protein in ^{13}C, ^{15}N enriched medium and purify to the level of >95%. The NMR sample should be degassed and filled with organ to avoid oxidation or paramagnetic effects of oxygen. As a part of solving NMR structure of protein it is essential to get sequential assignment for all atoms in the protein. To get sequential assignment, one has to get data of 2D ^{15}N-1H HSQC, CBCACONH, HNCACB, HBHAcbcacoNH, HNcaHA, hCCH-TOCSY on ^{13}C, ^{15}N enriched sample.

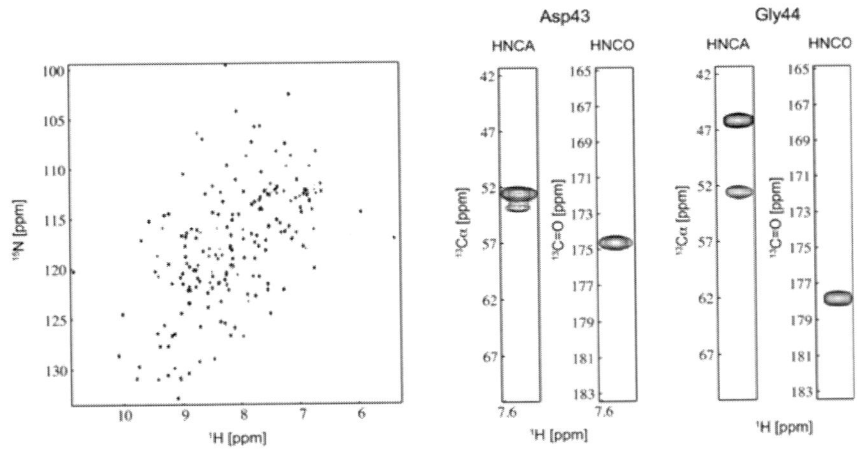

Figure 17. ^{15}N-^{1}H HSQC spectra of cellulose binding protein and strips used for sequential assignment.

The data would help you to make connectivity between adjacent amino acids by walking from N-terminal to C-terminal of a protein as shown in figure 17. As shown in flowchart in figure 16, once after completion of sequential assignment, the distance and conformational constraints between atoms and groups are derived using NOESY, EXSY experimental data. Finally, structure refinement was done using energy minimizing programs with the help of NMR derived distance constraints, coupling constants, residual dipolar coupling constants and conformational constraints.

REFERENCES

Abrgam A., *"The Principles of Nuclear Magnetism"* New York and London, Oxford University Press, 1961.

Branden C. and Tooze J., *"Introduction to Protein Structure"* 2nd Edition, New York, Garlanad publishing Inc., 1991.

Drenth, *"Principles of Protein X-ray Crystallography"* New York, Springer-Verlag, 1994.

Paul D. Stanely *"Principles and Topical application of 19F NMR Spectroscopy"* in Organofluorine-the hand book of environmental chemistry, Springer Verley Berlin Heidelberg, volume 3N, 2002 pp 1-61.

Vasle DINOIU, *"Fluorine Chemistry: Past, Present and Future"* Revue Roumaine de chimie, 2006, 51 (12) pp 1141-1152.

Wuthrich K., *"NMR of Proteins and Nucleic Acid"* New York, Wiley, 1986.

In: Nuclear Magnetic Resonance (NMR)
Editor: Doddapuneni Krishna Rao
ISBN: 978-1-63117-768-2
© 2014 Nova Science Publishers, Inc.

Chapter 2

STRUCTURAL CHARACTERIZATION OF FLUCONAZOLE DRUG USING NMR SPECTROSCOPY[†]

Sachin Kumar[1] and Doddapuneni Krishna Rao[2,††]

[1]Jubilant Chemsys Ltd. R and D Centre Noida, Noida, Uttar Pradesh, India
[2]Singapore Bioimaging Consortium, A*STAR, Singapore

ABSTRACT

The power of Nuclear Magnetic Resonance spectroscopy (NMR) in structure elucidation derives in large part from its ability to establish bonding connectivity (via J- coupling interaction) or through space proximity (via dipolar coupling interactions) of nuclei.

The amount of time consumed in elucidating a structure depends on the rate at which these interaction can be detected by NMR and analyzed.1D NMR methods most often explore interactions between only few nuclei at a time: spin-spin decoupling measurements are used to demonstrate through-bond connectivity; and NOE measurements are used to probe inter-nuclear distances, 2-Dimensional NMR experiments provide much more structural information in a given time period.

In this chapter we would like illustrate the NMR experiments used to find the drug molecule structure to the atomic level.

---

† Approved by Dr. Prasant Deb and Dr. Partha Mukherjee

†† Correspondence: kri_uoh@yahoo.co.in and bhullarsachin552@gmail.com.

Keywords: NMR, J-coupling, dipolar coupling, NOE, 2 dimensional NMR and 19F decoupled proton nmr

INTRODUCTION

Soon after the commercial NMR spectrometer available in 1960's, it is extensively used for solving the molecule structure to the atomic level using ^{1}H and ^{13}C NMR. It has become essential for all pharmaceutical industry to have NMR spectrometer to resolve structural issues of their synthesized molecules as new drugs.

In this prospect, we believe this chapter would be helpful for the reader to know how the drug molecule structure can be analyzed by using NMR spectroscopy. We choose an antifungal drug, Diflucan and Trican trading by Pfizer to illustrate the NMR methods for solving the new drug molecules.

Fluconazole is a triazole antifungal drug used in the treatment and prevention of superficial and systemic fungal infections. In a bulk powder form, it appears as a white crystalline powder, and it is very slightly soluble in water and soluble in alcohol. It is mainly marketed under the trade names Diflucan and Trican (Pfizer).

The discovery of Diflucan (fluconazole) was a major landmark in the pharmaceutical industry, as it was the first antifungal drug to be developed that could be used both orally for minor infections such as Candidiasis and intravenously for more serious systemic infection like cryptococccal meningitis.

Diflucan tablets are available in 50, 100, 150, and 200 mg of fluconazole and its formulation contains microcrystalline cellulose, dibasic calcium phosphate anhydrous, povidone, croscarmellose sodium, FD and C Red No. 40 aluminium lake dye, and magnesium.

Diflucan is also available in oral suspension with a dosage of 350 mg and 1400 mg with formulation of sucrose, sodium citrate dihydrate, citric acid anhydrous, sodium benzoate, titanium dioxide, colloidal silicon dioxide, xanthan gum, and natural orange flavour. Diflucan is available for form of injection with 2mg/ml with formulation of sodium chloride 9mg/ml or 56mg/ml of dextrose.

Mechanism of Action: Like other imidazole and triazole class of antifungals, fluconazole inhibits the fungal cytochrome P450 enzyme 14α-demethylase [1]. Mammalian demethylase activity is much less sensitive to fluconazole than fungal demethylase [2].

The inhibition of demethylase prevents the conversion of lanosterol to ergosterol, which is an essential component of the fungal cytoplasmic membrane, and subsequent accumulation of 14α-methyl sterols.

Fluconazole is primarily fungistatic; however, it may be fungicidal against certain organisms in a dose-dependent manner, specifically *Cryptococcus* [3].

FLUCONAZOLE IS EFFECTIVE AGAINST THE FOLLOWING FUNGAL INFECTIONS

- Oesophageal and oropharyngeal *candidiasis*
- Vaginal *candidiasis* due to *Candida* yeast infections
- *Cryptococcal meningitis,* especially in AIDS patients
- Urinary tract infections arising from *Candida* infections
- Peritonitis
- Systemic *Candida* infections including *candiedmia,* disseminated *candidiasis* and pneumonia
- Prophylaxis especially for patients undergoing bone marrow transplantation who receive cytotoxic chemotherapy and/or radiation therapy.

Resistance: Fungal resistance to drugs in the azole class tends to occur gradually over the course of prolonged drug therapy, resulting in clinical failure in immunocompromised patients [4] (e.g., patients with advanced HIV receiving treatment for thrush or esophageal *Candida* infection).

In *C. albicans*, resistance occurs by way of mutations in the *ERG11* gene, which codes for 14α-demethylase. These mutations prevent the azole drug from binding, while still allowing binding of the enzyme's natural substrate, lanosterol.

Development of resistance to one azole in this way will confer resistance to all drugs in the class. Another resistance mechanism employed by both *C. albicans* and *C. glabrata* is increasing the rate of efflux of the azole drug from the cell, by both ATP-binding cassette and major facilitator superfamily transporters.

Other gene mutations are also known to contribute to development of resistance [5].

EXPERIMENTAL SECTION

Synthesis: Before studying the structure of Fluconazole, it is essential to know the synthesis route so that you will get preliminary information about structure and functional groups. Fluconazole can be synthesized from a halogenated acetophenone derivative as shown in Figure 1.

NMR sample preparation: The Fluconazole compound was synthesized and purified to greater than 95% and dried completely to remove the solvents from synthesis and purification step.

20-30 mg synthesized sample or the commercially available tablet(s) was dissolved in 500uL of DMSO-d6 and transferred the clear solution into 5mm NMR tube. The filled NMR tube is placed in the 400 MHz Bruker magnet with 5mm BBFO probe. The magnet is locked to deuterium signal of DMSO, to avoid the drift in the magnetic field. The magnetic field was homogenized by shimming the coils around probe and sample to achieve better line shape and sensitivity. The ^{1}H pulse was calibrated using solvent resonances. ^{1}H, ^{13}C-APT, HMQC, HMBC, and ^{19}F experiments were performed using the bruker library of experiments after adjusting the acquisition parameters.

RESULTS AND DISCUSSION

To solve the structure of Fluconazole drug, we have used the ^{1}H NMR, ^{13}C NMR (APT or DEPT), ^{13}C-^{1}H HMQC, HMBC, and ^{19}F NMR. As discussed in the chapter 2 in this book, ^{1}H NMR and ^{13}C NMR is used to identify the number and type of protons and carbons in the structure. ^{13}C-^{1}H HMQC is used to know for which proton is directly connected to which carbon. ^{13}C-^{1}H HMBC is known to produce ^{2}J, ^{3}J, and even sometimes ^{4}J couplings between carbon and proton, these correlations are useful in making connectivity's between carbon-carbon and carbon-protons.

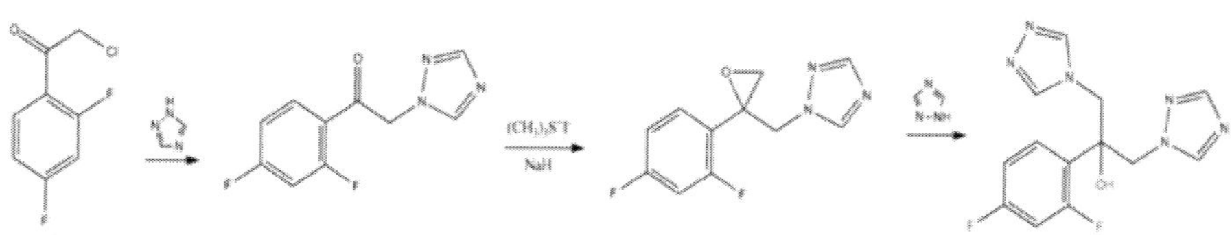

Figure 1. Fluconazole synthetic route [6].

Figure 2 showing the ^{1}H NMR spectrum of Fluconazole drug in DMSO-d6 solvent recorded using 400 MHz NMR spectrometer. The ^{1}H NMR spectrum integration suggests that this drug structure contains 12 protons. Based on the chemical shift range, out of those 12 protons 7 protons are resonating in the aromatic region, with this we might consider there are 7 aromatic protons. The D_2O exchange spectrum suggests that there is one exchangeable proton in this molecule. In the aliphatic region two protons of each showing geminal coupling of about 14 Hz, although those two protons on the same carbon, they were coupled to each other due to diastereotopic nature of protons. In some cases we can predict connectivity between protons by looking at the splitting pattern, where as in this case splitting pattern is complex. The complex splitting pattern might be due to other nuclei like Fluorine or other magnetically active nuclei. This can be clarified by doing ^{19}F NMR or ^{19}F decoupled ^{1}H NMR. To resolve the complex splitting pattern from 6.8-6.9 and 7.1-7.3 ppm region, we performed a ^{19}F decoupled ^{1}H spectrum, which can suppress the ^{19}F coupling to the ^{1}H. As shown in the figure 2, in the case ^{19}F decoupled spectrum clearly suggest that the proton in the region of 6.8-6.9 ppm is having coupling with both ortho and meta positioned protons. The two protons in the spectral region of 7.1-7.3 ppm one proton shows a coupling of 8 Hz ((7.17-7.15)*400) which indicates it is ^{3}J-ortho coupling of proton and the other proton is showing coupling of 3-4 Hz suggests ^{4}J-meta-coupling of a proton.

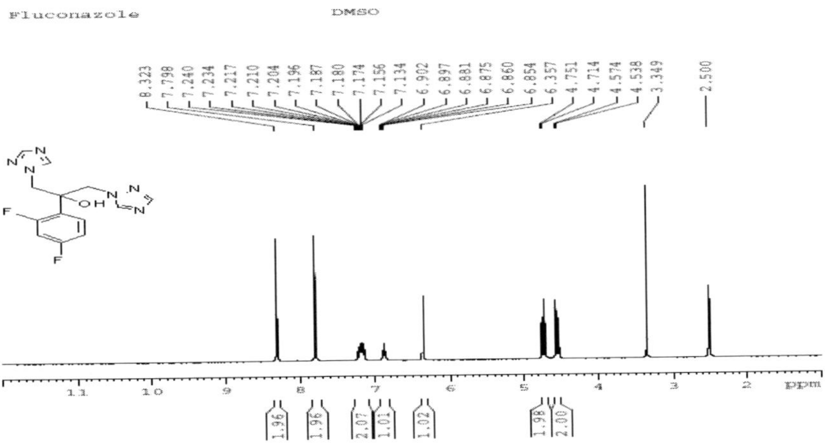

a

Figure 2. (Continued)

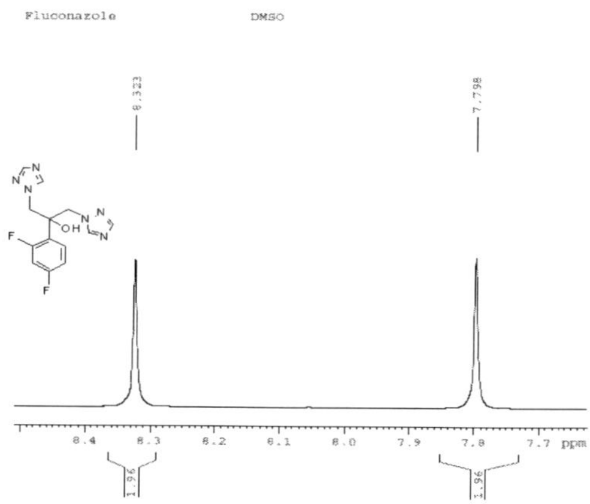

b

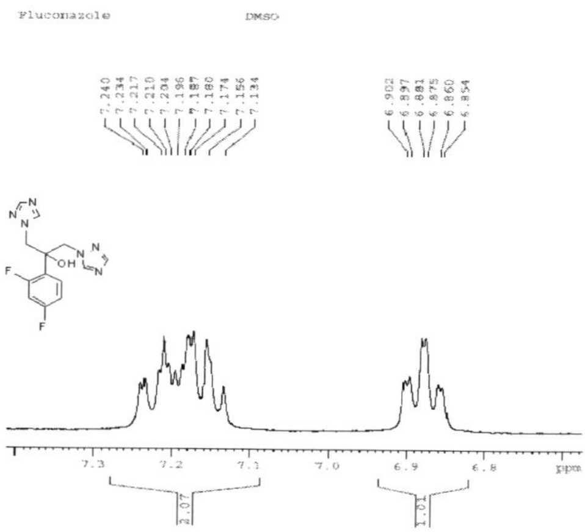

c

Figure 2. ^{1}H NMR spectrum of Fluconazole in DMSO-d6. The expanded spectra of aromatic region is shown in (b) and (c).

In figure 3, the ^{19}F NMR clearly indicates that there are two fluorine atoms in the molecule; the coupling constant between two fluorine atoms is less than 15 Hz suggest that those fluorines are either meta or para to each other on Benzene ring.

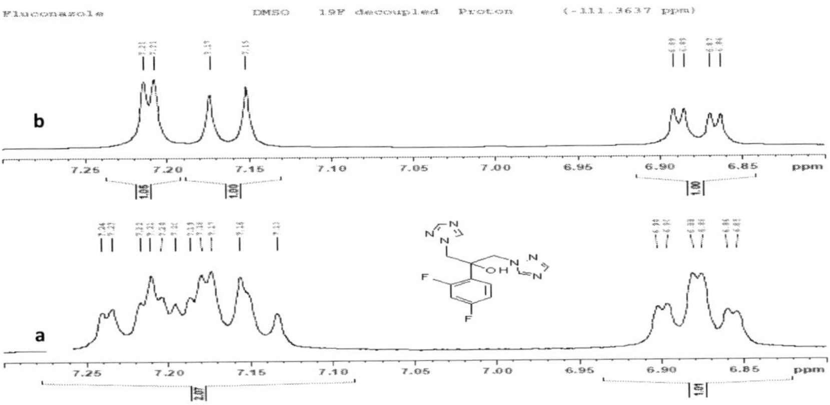

Figure 3. The overlay of aromatic region of ^{19}F coupled (a) and ^{19}F decoupled (b) ^{1}H NMR spectra of Fluconazole.

Although the ^{1}H and ^{19}F decoupled NMR gives clear information about splitting pattern, it is essential to run ^{13}C NMR to detect the connectivity between atoms in the molecule. To discover the connectivity, we performed attached proton test (^{13}C-APT), HMQC, and HMBC experiments.

One of the useful way of determining how many protons attached carbon in a molecule can be identified using an attached proton test (APT), which distinguishes between carbon atoms with even or odd number of attached protons.

As shown in figure 4, the APT spectrum reports that there are 4 quaternary carbons, 5 single proton attached carbons, and 2 two proton attached carbons. Based on the ^{1}H integration there are 4 symmetric protons in the structure, two in the aliphatic and two in the aromatic region. By looking at HMQC (figure 5) and APT spectra, the total number of carbons is equal to 15. With hetero nuclei multi quantum correlation (HMQC) spectrum, we can understand the direct connectivity between carbon and proton. When we look back to APT spectrum, the quaternary carbons shows doublet of doublet, this coupling is mainly due to ^{1}J (290 Hz) and ^{3}J (12 Hz) carbon-fluorine. The coupling pattern supports that these two fluorine atoms are meta (1,3 position) to each other on benzene ring. Although by now we are able to recognize number of carbons and protons and their direct attachments, we still need to make connectivity to make complete molecule structure. Heteronuclear multiple bond correlation (HMBC) is performed to make connectivity between atoms in the molecule using long rang correlations.

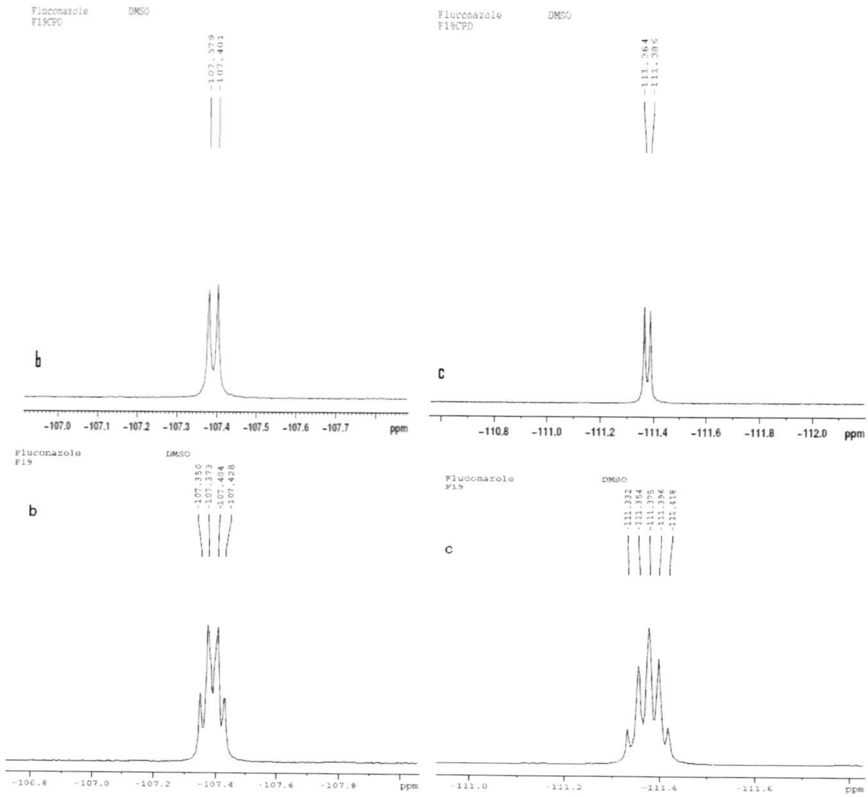

Figure 4. ^{19}F NMR spectra with ^{1}H coupled and decoupled.

In this experiment all ^{1}J (one bond) couplings are suppressed, where as HMQC shows only one bond coupling. Based on coupling constants and delay time, we can extend correlations up to 3 or 4 bonds.

As shown in the figure 6, we have acquired HMBC data on fluconazole sample on 400 MHz NMR spectrometer. We do see the following correlation in our spectrum. We observed a correlation between aliphatic quaternary carbon at 54.91 ppm with OH proton at 6.3 ppm and benzene ring aromatic CH at 7.1 ppm and both aliphatic CH2 at 4.7 and 4.5 ppm. The carbon at 55 ppm correlated with OH proton at 6.3 ppm and CH_2 protons at 4.7 and 4.5 ppm. The carbon at 55 ppm supposed to show a correlation with pyrazole ring protons, but due to nitrogen interference the correlation might get suppressed.

The quaternary carbon resonating at 73.63-73.68 ppm is showing HMBC correlations with the ^{1}H resonating at 7.15, 6.3, 4.5 and 4.7 ppm. As shown in figure 7 and table 1, all the assignments and correlations can be validated using HMBC and HMQC data.

CONCLUSION

This chapter describes the route map of structural elucidation of drug molecules using NMR. The same approach can be used to elucidate new drug molecule structure. In this particular example, we have used ^{19}F decoupled ^{1}H spectra to explain the complex splitting patterns in 1H NMR. This example helps the reader to use how ^{19}F can be used to make structural elucidation much simpler. Based on ^{19}F coupling constants one can identify the position of two fluorines on aromatic ring. Although the HMBC, HMQC gave routine information, these two experiments are critical to explain the connectivity between atoms.

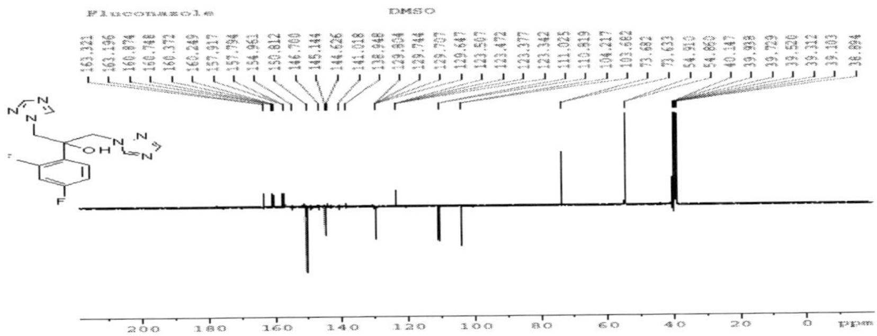

a

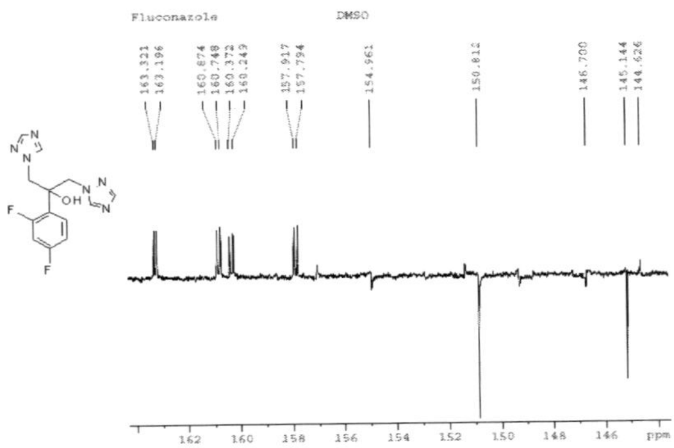

b

Figure 5. (Continued)

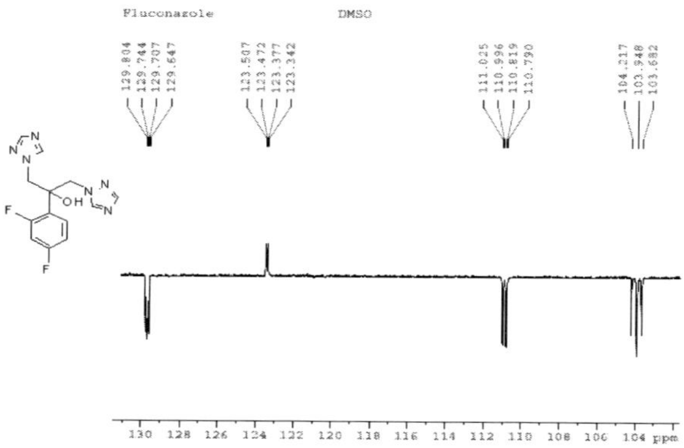

c

Figure 5. [13]C Attached proton test (APT) NMR of Fluconazole. The full spectrum (a) and expanded aromatic region (b and c).

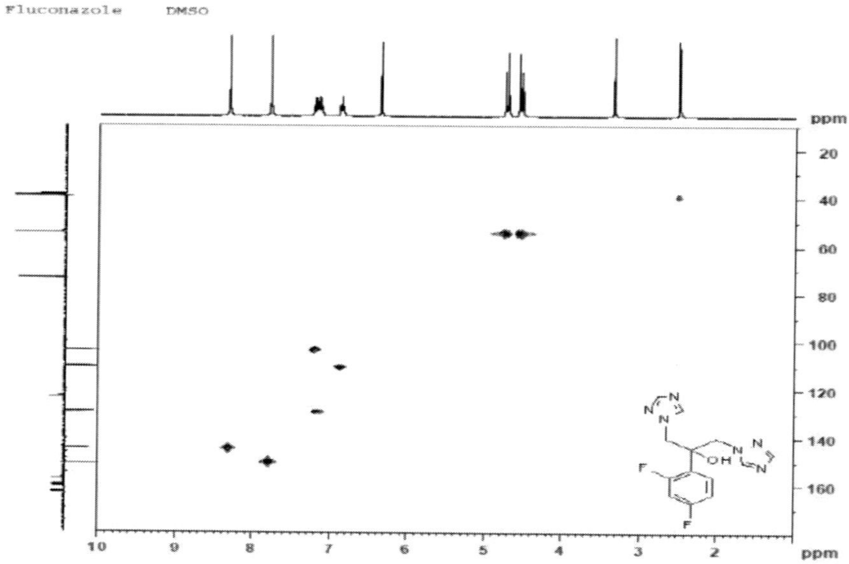

a

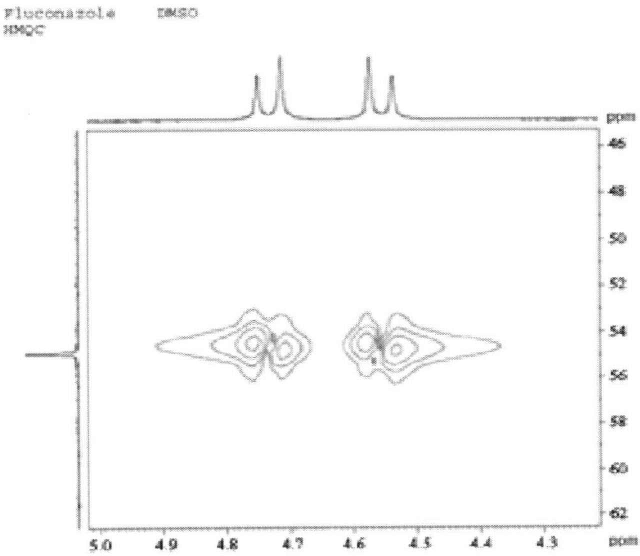

b

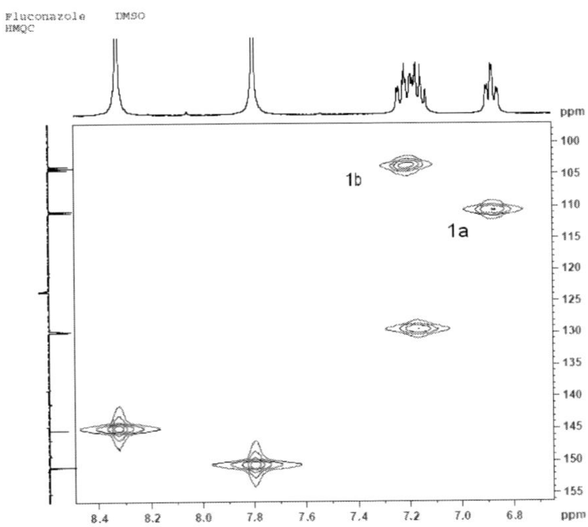

c

Figure 6. ^{1}H-^{13}C HMQC spectrum of Fluconazole. ^{13}C-APT and ^{1}H NMR spectra were plotted as projections on both dimensions.

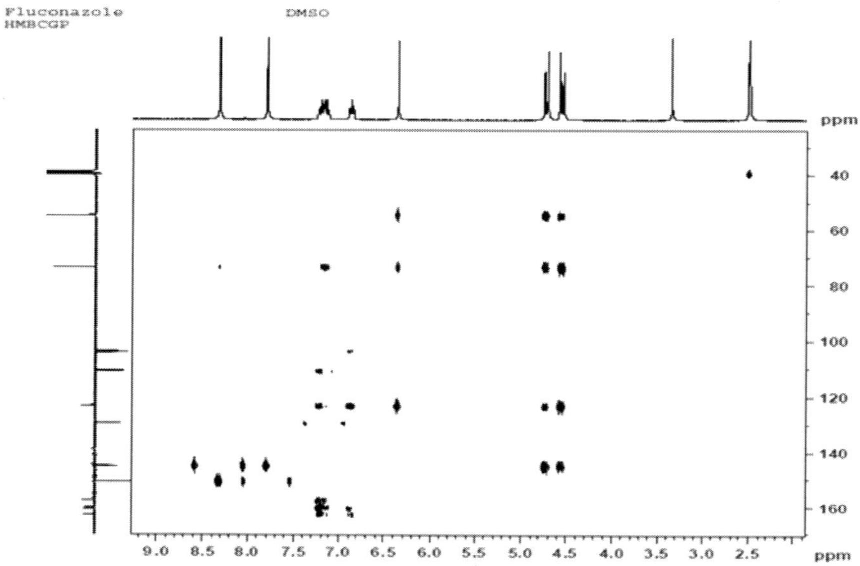

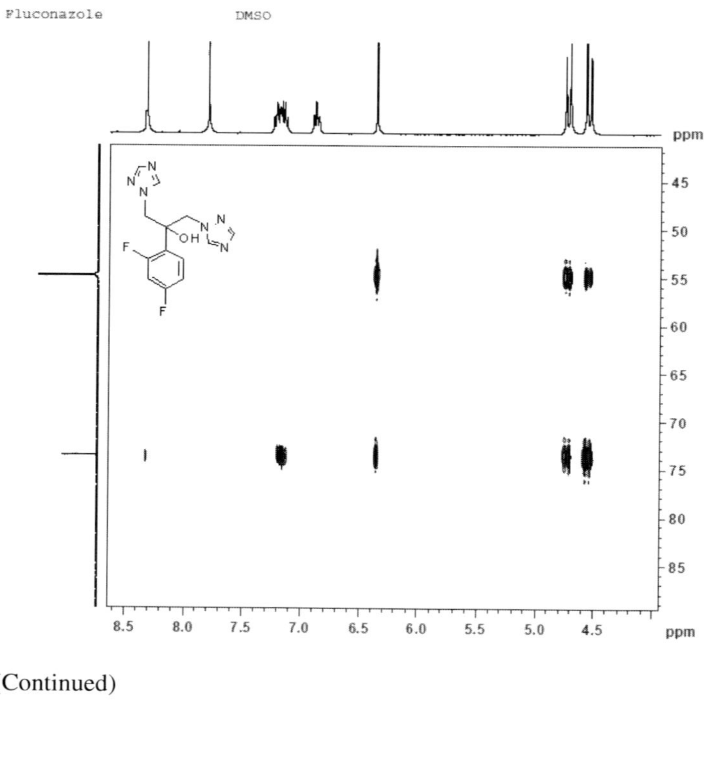

Figure 7. (Continued)

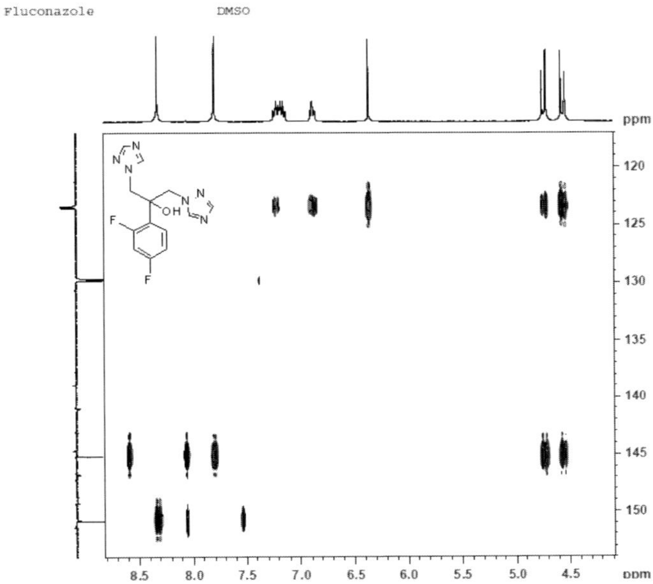

Figure 7. ^{13}C-^{1}H HMBC spectra of fluconazole with expanded regions.

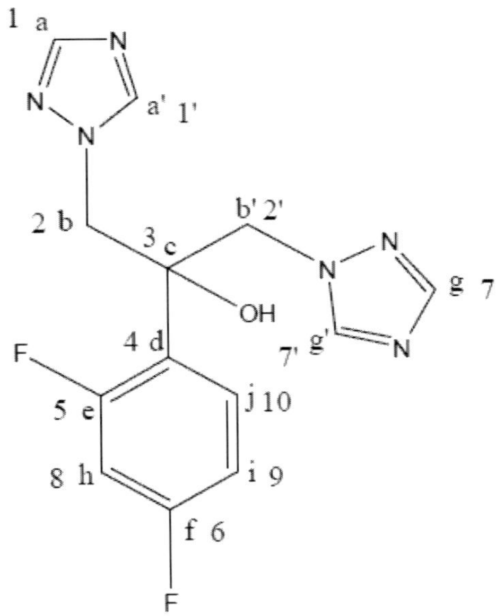

Figure 8. The final structure of Fluconazole (2-(2,4-difluorophenyl)-1,3-di(1H-1,2,4-triazol-1-yl)propan-2-ol).

Table 1. The chemical shift values, type and J-couplings of ^1H and ^{13}C

S.No.	Name	^1H NMR shift	^{13}C NMR shift	Name	Nature of Carbon
1.	1'	8.323ppm;s	a' 145ppm	a'	CH
2.	1	7.798ppm;s	a 151ppm	a	CH
3.	2	4.751ppm;d	b 55ppm	b	CH_2
4.	2'	4.574ppm;d	b'55ppm	b'	CH_2
5.	3	Quaternary	c (73.682,73.633) (3J with 19 F) ppm;d	c	C
6.	4	Quaternary	d 123.50;dd	d	C
7.	5	Quaternary	e 163,160.8ppm;d	e	C
8.	6	Quaternary	f 160.3,157.9ppm;d	f	C
9.	7'	8.323ppm;s	g 145ppm	g'	CH
10.	7	7.798ppm;s	g' 151ppm	g	CH
11.	8	(7.214,7.208) ppm;d (3j)	h (104.217 103.682) ppm;t	h	CH
12.	9	(6.892, 6.886) (4j) (6.892,6.864) (3j) ppm;dd	I (111.996 110.819) ppm;dd	i	CH
13.	10	(7.174, 7.152) (3j) ppm;d	j (129.804 129.647) ppm;dd	j	CH
14.	OH	6.3ppm;s	No carbon	OH	-

REFERENCES

[1] Vanden Bossche, H., Dromer, F., Improvisi, I., Lozano-Chiu, M., Rex, J. H., Sanglard, D. (1998). "Antifungal drug resistance in pathogenic fungi". *Med. Mycol.* 36, 119–28.

[2] Mullins, J. G. L., Parker, J. E., Cools, H. J., Togawa, R. C., Lucas, J. A., et al. (2011). "Molecular Modelling of the Emergence of Azole Resistance in *Mycosphaerella graminicola*. *PLoS ONE* 6(6): e20973.

[3] Longley, Nicky; Muzoora, Conrad; Taseera, Kabanda; Mwesigye, James; Rwebembera, Joselyne; Chakera, Ali; Wall, Emma; Andia, Irene et al. (2008). "Dose Response Effect of High-Dose Fluconazole for HIV-Associated Cryptococcal Meningitis in Southwestern Uganda". *Clinical Infectious Diseases* 47 (12): 1556–1561.

[4] Sweetman, S., editor. *Martindale: The complete drug reference*. 34th ed. London: Pharmaceutical Press; 2004. ISBN 0-85369-550-4.

[5] Bennett, J. E. (2011). *Chapter 57. Antifungal Agents*. In: L. L. Brunton, B. A. Chabner, B. C. Knollmann (Eds), Goodman and Gilman's The Pharmacological Basis of Therapeutics.

[6] Richardson, R. K.; 1983, US Patent 4,404,216.

In: Nuclear Magnetic Resonance (NMR) ISBN: 978-1-63117-768-2
Editor: Doddapuneni Krishna Rao © 2014 Nova Science Publishers, Inc.

Chapter 3

OFF-PATHWAY INTERMEDIATE(S) IN THE REFOLDING OF ALKALINE FERROCYTOCHROME *C*: A PULSE-LABELING HX NMR STUDY

***Doddapuneni Krishna Rao,*[1,*]**
***N. Prakash Prabhu*[2] *and Abani K. Bhuyan*[3]**
[1]Singapore BioImaging Consertum, A*STAR, Singapore
[2]School of Life sciences, University of Hyderabad, India
[3]School of Chemistry, University of Hyderabad, India

1. ABSTRACT

Proteins meet with the stipulations of Levinthal. The classical mechanism involving a pre-transition search topology step can explain adequately how the biologically relevant time scale is achieved. Association of the classical mechanism of chemical kinetics naturally allows depiction of folding pathways with dead-end intermediates. By using the approach of hydrogen exchange (HX) pulse labeling in conjunction with NMR spectroscopy, it is shown here that carbonmonoxycytochrome *c* when allowed to refold from a strongly alkaline medium to neutral pH folds via a highly misfolded state. The misfolded state placed at the dead end of the folding pathway must then

---

* E-mail: kri_uoh@yahoo.co.in.

melt to the initial unfolded state to have a chance to refold to the correct native state.

2. INTRODUCTION

Emerging ideas of landscape perspective or the funnel model [1-6] were first used to interpret both two-state folding kinetics and accumulation of misfolded structures on the basis of size, slope, and roughness of energy landscapes [7, 8]. Corroboration between experimental and simulation results has also been shown regarding the fractional φ values for CI2 [9]. The landscape view holds that the funneled organization of the energy landscape dominates the folding kinetics, because the great variety of detailed mechanisms existing on the funnel influences the folding trajectories [6]. The emerging ideas, however, often appear to undermine some of the strongly established empirical patterns of folding kinetics. These patterns, including the energy and the location of the folding barrier, that form the key elements of the classical view of folding, are not reproduced in the new paradigm. Overlooking empirical results is difficult. A number of recent studies have demonstrated the involvement of sequential intermediates in the folding pathway of cyt *c* [10-13]. It has also been shown that the misfold organization of the folding intermediate is associated with an optional event, and it does not negate the classical description of folding [1]. Application of the classical paradigm to a number of proteins exhibiting diverse folding behavior has been discussed [14].

Supported by empirical observations on ferrocytochrome *c* (ferrocyt *c*), this work shows extreme misfolding when alkaline ferrocytochrome *c* liganded with CO is allowed to refold to a final pH of 7. Earlier reports of misfold organization of a kinetic intermediate of ferricytochrome *c* (ferricyt *c*) also raised the question of how conducive intermediates are to folding [15, 16]. Whether it is conducive or not, polypeptides have the option to make mistake and misfold. If a misfolded state is blocked from further folding, it has to go back to the initial unfolded state to start afresh. Results presented here elucidate this situation.

3. Materials and Methods

Cyt *c* was from Sigma (Type VI). GdnHCl and sodium dithionite were obtained from USB and Merck, respectively. Experiments were done in strictly anaerobic atmosphere at 22°C using 0.1 M sodium phosphate buffer for pH 7 conditions, and NaOH or NaOD (with or without 1 mM CAPS) for pH 12.7 conditions. Solutions contained 0.5-3 mM freshly prepared sodium dithionite, and experiments were completed within 2 hours of exposing the protein to high pH. High-pH fluorescence changes were corrected by using NATA fluorescence.

3.1. Equilibrium Unfolding

Cyt *c* solutions with or without 1mM CO, prepared in the 0-7 M range of GdnHCl or 7-13.25 range of pH, were deaerated and reduced under nitrogen with ~2 mM sodium dithionite, and incubated in tightly capped quartz cuvettes or rubber-capped small glass tubes for ~30 minutes. Tryptophan fluorescence emission spectra (ex: 280 nm) were taken in a FluoroMax-3 instrument (Jobin-Yvon, Horiba). Optical absorption spectra were recorded in a Cary 100 (Varian) spectrophotometer. Data were analyzed using the standard two-state equations [17].

3.2. Stopped-Flow Kinetics

Cyt *c* initially dissolved at pH 13 in the presence or absence of 4 M GdnHCl, was reduced under nitrogen by adding a concentrated solution of sodium dithionate to a final concentration of ~3mM. The solution was then saturated with CO under 1 atm of the gas. The resulting carbonmonoxycyt *c* is unfolded under the conditions used. Refolding was initiated by eight-fold dilution of the protein solution with the refolding buffer. The final protein concentration in the refolding mixture was in 5-50 μM range. For unfolding experiments, the native state of carbonmonoxycyt *c* (NCO) was prepared by manual dilution of the alkali-unfolded protein with the refolding buffer in 1:40 ratio. NCO was then unfolded by eight-fold dilution into the unfolding buffer (0.1 M phosphate, 3 mM sodium dithionite, pH 7) containing different concentrations of GdnHCl. The final protein concentration in these experiments was ~4 μM. A Bio-Logic SFM 400 instrument was used for these experiments.

3.3. Hydrogen Exchange Pulse Labeling and NMR Spectroscopy

Cyt *c* with fully deuterated backbone amide sites was dissolved in NaOD at pD 12.9 with or without 3 M GdnDCl, reduced with sodium dithionite (3 mM), and unfolded by adding 1 atm CO. Initial protein concentration was 6 mM. Refolding was initiated by 6-fold dilution into a H_2O buffer containing 75 mM phosphate, pH 6.4. The final pH of the refolding medium was 7.3. After variable times of refolding, the solution was mixed with an equal volume of the H_2O pulse buffer containing 50 mM CAPS, pH 10.4. The pH of the mixture at this stage was 9.6. After 50 ms of pulse time, the solution was combined with 0.7 volume of an H_2O quench buffer containing 0.5 M sodium acetate and 0.35 M ascorbate, pH 5. The pH at this stage was 5.05. The quenched solution was washed and concentrated at 5°C using a D_2O buffer consisting of 7 mM citrate and 7 mM ascorbate, pH 5. The samples were then subjected to NMR analysis by recording magnitude COSY and TOCSY spectra in 400 MHz (Bruker) or 600 MHz (Varian) spectrometers. Proton occupancies at individual amide sites were calculated by using the cross peak intensities of a control unfolded sample whose amide sites were labeled maximally. Rates were extracted from plots of proton occupancies as a function of refolding time. Pulse-labeled samples were prepared using the Bio-Logic instrument in the quench-flow mode.

For samples not subjected to pulse labeling, D_2O solutions of 1 mM cyt *c*, the amides of which were preexchanged, were adjusted to pH 7 (0.1 M phosphate) or pH 12.7 using NaOD. Unfolded protein was prepared in GdnDCl. For reduction solid sodium dithionite was added under nitrogen to obtain a final concentration of 20 mM. The tubes were sealed with sleeved rubber stoppers. Spectra were taken at 22°C in a 400 MHz Bruker spectrometer (AV400).

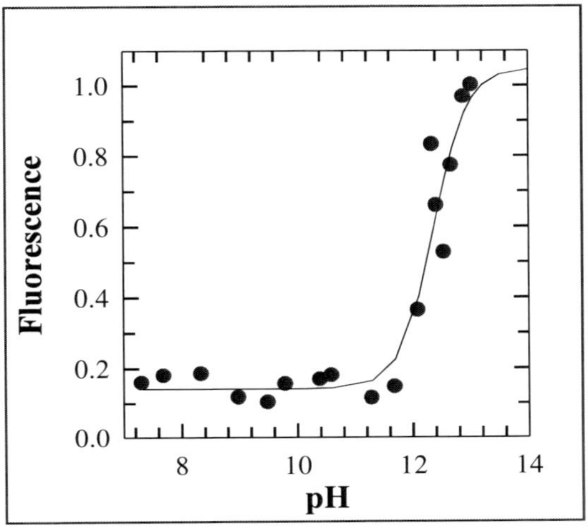

Figure 1. The alkaline pH-induced unfolding of ferrocyt *c* in the presence CO. The pH-midpoint for the transition is ~12.5.

4. Results

Ferrocytochrome *c* (cyt *c*) unfolds in alkaline aqueous medium when carbonmonoxide is allowed to bind to its ferrous heme iron (Figure 1).

The unfolding of the resultant CO-bound protein, called carbonmonoxycytochrome *c* (Cyt-CO) is substantial even in the aqueous medium, because the addition of the denaturant guanidinium hydrochloride (GdnHCl) to the protein solution at pH 13 or higher does not produce any considerable change in the fluorescence emission by lone tryptophan (W59). This however suggests little that the alkaline Cyt-CO is random coil-like. Completeness of unfolding is a controversial issue and cannot be shown unequivocally with a number of techniques.

4.1. Neutral and Alkaline Forms of Ferrocyt *c*

On the other hand, ferrocyt *c* at pH 12.7 (alkaline ferrocyt *c*) is substantially structured. The NMR spectrum in Figure 2a shows chemical shift dispersion characteristic of a secondary- and tertiary-structured protein. Minor changes in both intensities and chemical shifts of resonances, relative to those seen for the native protein at pH 7 (neutral ferrocyt *c*), indicate some readjustments of side chains. The spectrum does not hint at molecular aggregation. In fact, TOF mass spectra and extensive ligand- binding experiments (data not shown) indicate no deamidation or aggregation at alkaline pH. A large number of NOE peaks seen at neutral pH appear at alkaline pH also, albeit somewhat shifted (Figure 2b, c).

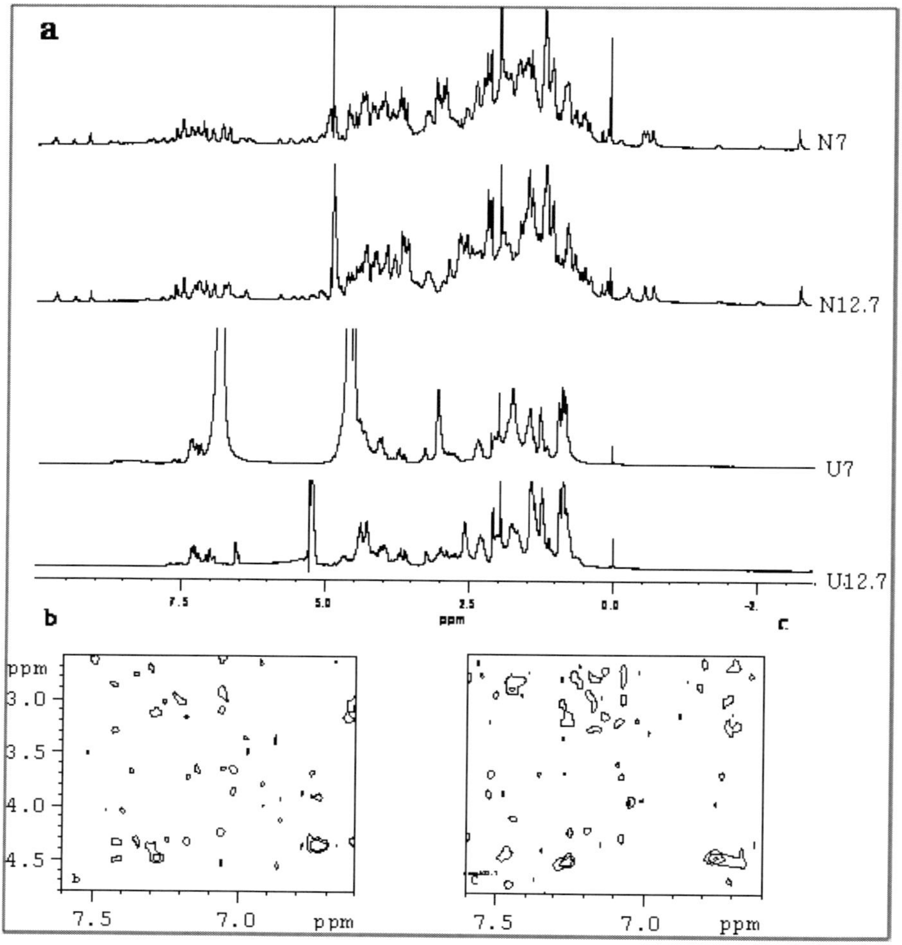

Figure 2. NMR spectra of ferrocytochrome *c*. (a) Native states at pH 7 (N) and pH 12.7 (N12.7), and GdnDCl-unfolded states at pH 7 (U 7) and pH 12.7 (U112.7). (b and c) Regions native-state NOESY spectra of neutral and alkaline ferrocyt *c*.

Figure 2a also compares the GdnDCl-unfolded NMR spectra of neutral- and alkaline ferrocyt *c*. Chemical shift dispersion is lost in both, but differences between the two are apparent, indicating differences in unfolded chain configuration. The latter also shows base-catalyzed fast exchange of residual hydrogens of GdnDCl with residual water protons coalesce as the exchange rate approaches the difference of the two chemical shifts.

4.2. Stability of Neutral and Alkaline Ferrocyt *c*

Figure 3a compares GdnHCl unfolding transitions of the two forms. Two-state analysis of data yields a $\Delta G_D°$ value of 18(±0.3) kcal mol^{-1} and 3(±0.5) kcal mol^{-1} for neutral and alkaline ferrocyt *c* respectively. Such a large difference in $\Delta G_D°$ for the neutral and alkaline ferrocyt *c* indicates drastic reduction of aqueous stability of the latter. The equilibrium *m*-values, m_g, for the neutral and alkaline forms are 3.6(±0.1) and 1.2(±0.1) kcal mol^{-1} M^{-1}, respectively. Since the native-state NMR spectra indicate that significant portions of secondary and tertiary structures are similar in neutral and alkaline ferrocyt *c*, the observed disparity in the m_g value is likely due to relatively small solvent accessible surface area in the denatured state of the alkaline form.

4.3. Folding and Unfolding Kinetics

In general, kinetic traces were fitted to 1-exponential. In some cases 2-exponentials were needed to improve the overall quality of fits. The slow minor phase, the amplitude of which averages to ~10% of the observed signal, represents most likely the fraction of the oxidized protein. Figure 3b shows denaturant dependences of the logarithm of the fast refolding (unfolding) phase for the two proteins. In each case, the relaxation minimum closely matches the midpoint of the individual equilibrium unfolding transition (Figure 3a).

4.4. Chevron Inversion in Strongly Native Like Condition

Curiously enough, when Cyt-CO is allowed to refold from pH 13 to pH 7 with or without 4 M GdnHCl in the intial unfolded solution, the measured millisecond refolding rate constant under strongly native like conditions (<2 M GdnHCl) decreases linearly with decreasing concentrations of denaturant, producing a sharp inversion in the folding limp of the chevron (Figure 3c).

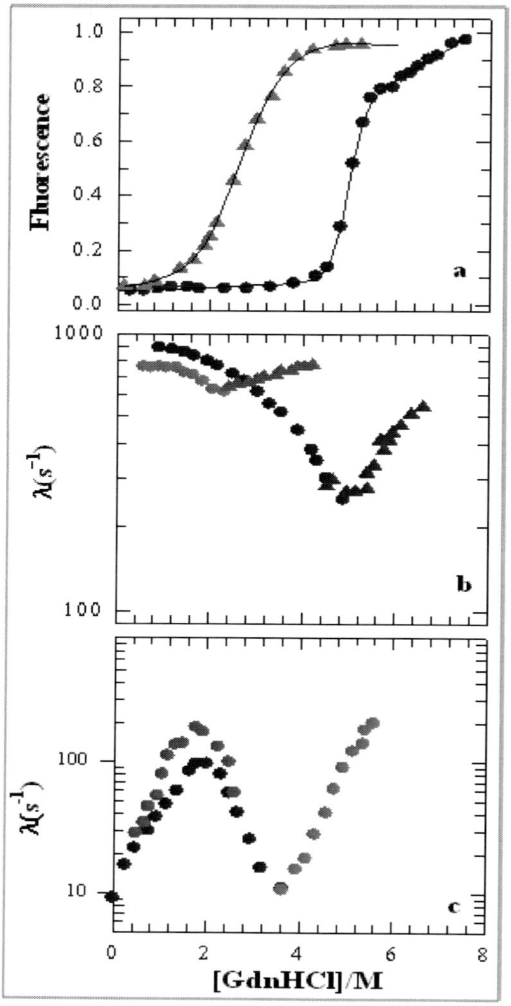

Figure 3. (a) GdnHCl unfolding of ferrocytochrome c at pH 7 (●), and pH 12.7 (▲). (b) GdnHCl dependence of rates for the pH 7 → pH 7 (●, folding and ▲, unfolding) and pH 12.7 → pH12.7 (●, folding and ▲, unfolding) equilibria. (c) Observed relaxation rate constants versus GdnHCl. The symbols (●) and (●) represent refolding rates when the alkali-unfolded CO-bound protein solution contained 0 M and 4 M GdnHCl, respectively. (●) represents the unfolding rates.

The accentuated inversion of the rate-denaturant profile is taken as an evidence for extensive misfolding: the misfolded species is increasingly stabilized and hence accumulates as the denaturant concentration in the refolding medium falls lower. The misfolded product cannot, however, achieve forward folding because of the preponderance of non-native interactions. In this sense it is a dead end intermediate, and it needs to unfold extensively in order to find a native like topology which can facilitate correct folding.

Here, it is necessary to show that the observed misfolding is not related to protein dimerization or aggregation at any stage of folding, since short-lived aggregates may cause a downward curvature in the folding limb of the chevron plot [18]. Protein aggregation at extreme pH was checked by several methods, including NMR and optical absorption (data not shown). Since, the sharp and well resolved resonances of the NMR spectrum of alkaline ferrocyt *c* (pH 12.9) do not change at all even after tens of hours of sample preparation (figure 4), the evidence is compelling that the protein is monomeric all the time.

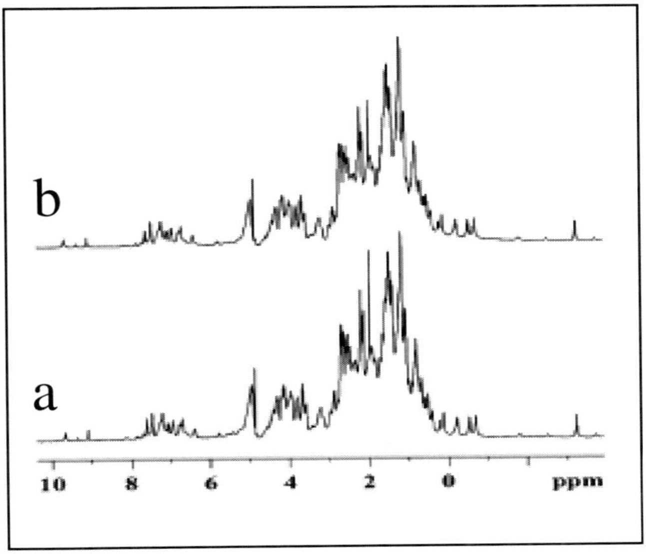

Figure 4. The protein does not aggregate or dimerize at different stages of folding. The nmr spectrum of ferrocyt *c* at pH 12.9 soon after sample preparation (a), and 38 h later (b).

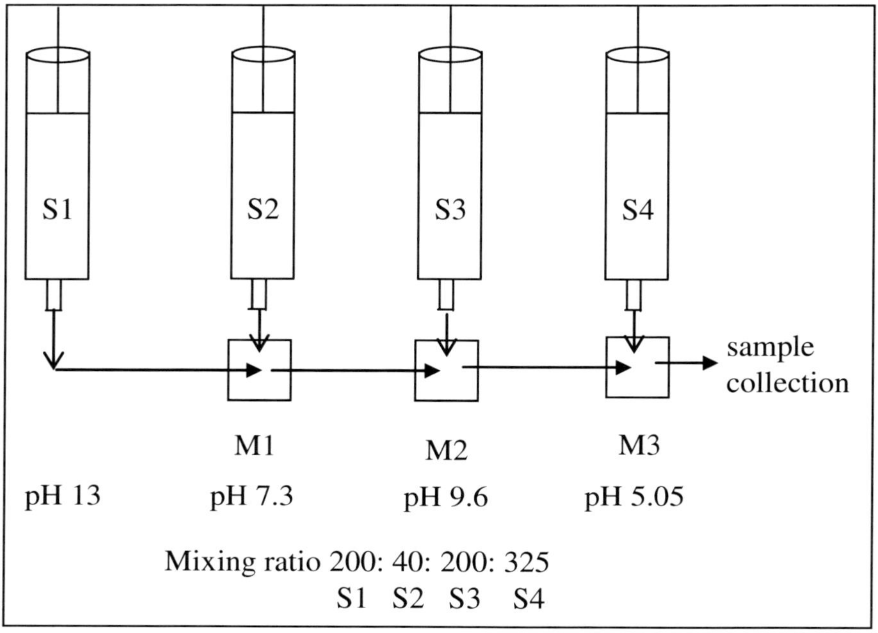

Figure 5. Schematic representation of the mixing sequence for the hydrogen exchange pulse labeling experiment. The mixing ratio was designed to achieve the pH values at different stages indicated.

4.5. Pulse Labeled Hydrogen Exchange Kinetics

The misfolding is better shown by using the pulse labeled hydrogen exchange strategy for probing structure formation during refolding. A Cyt-CO solution unfolded in the alkaline medium in the presence of GdnHCl was allowed to refold to a final condition of pH 7 and 0.5 M GdnHCl using the quench-flow protocol as shown in Figure 5. The pH of the solutions at different stages of refolding and the delay-times (aging time) were standardized to maximize labeling during refolding but to minimize exchange in the final refolded state. Control samples to check for pulse leakage were prepared by manual mixing instead of using the quench-flow mixing. Both TOCSY and COSY spectra were run on the concentrated and washed samples. Figure 6 shows the NH fingerprint regions of representative COSY spectra depicting the H/D exchange as folding progresses.

For all NMR-resolved NH sites, only moderate protection against proton labeling was seen at ~14 ms of refolding time, although at 55ms the proton occupancy dropped to about 60-70% in a lone kinetic phase corresponding to the $I \rightarrow N$ transition via U.

The rate extracted from the time course of protection against H-exchange labeling at four representative NH sites is 30 ± 2 s^{-1} (figure 6D). This rate is also seen in stopped-flow interrupted refolding experiments (data not shown). The single rate for protection suggests that no transient structure accumulates during the folding.

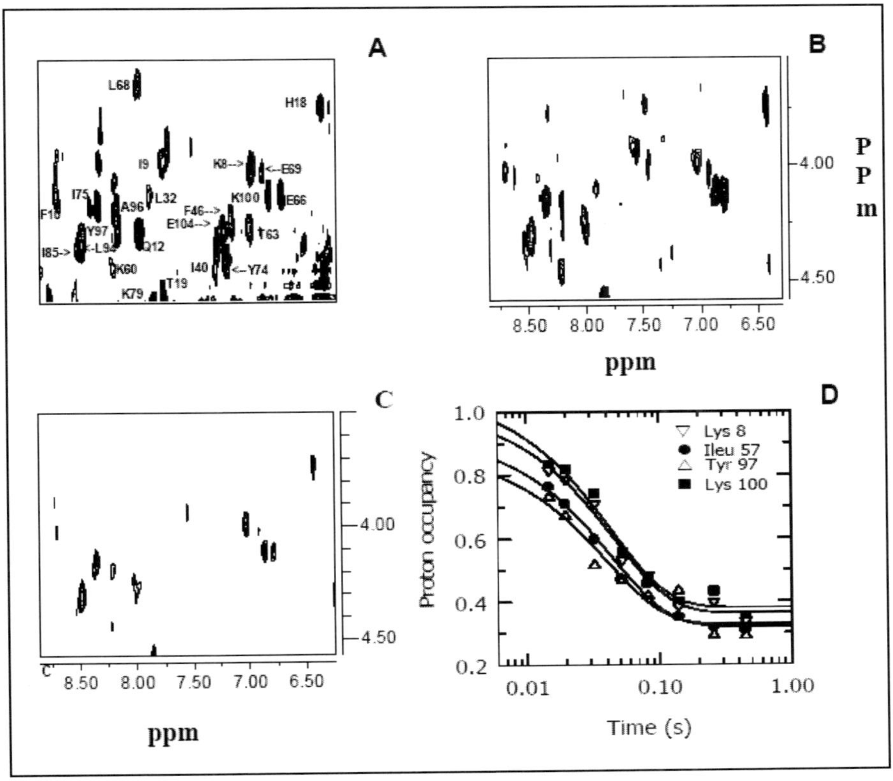

Figure 6. NMR analysis of representative samples prepared by hydrogen exchange pulse labeling. (*A*) NH-C_{α}H region of the magnitude COSY (*J*-correlated) spectrum of ferrocyt *c* at pH 5.1. (*B*) As in (*A*) for a pulse-labeled sample at 14 ms of refolding time. (*C*) As in (*B*) at 55 ms of refolding time. (*D*) Time course of protection against hydrogen exchange for a set of four representative amide sites. Single-exponential fits to the data give 30 ± 2 s^{-1} for the refolding rate constant.

5. Discussion

5.1. Folding Models and the Reality

The principles and the relaxation mechanisms by which the unfolded polypeptide acquires the native state are not understood. There are models, classical and contemporary, for the description of the folding phenomenon. While the classicals are deeply rooted into the principles of chemical kinetics and the law of mass action, the contemporaries are fundamentally based on the principles of statistical mechanics. It must be kept in mind though that a model, whatever is the basis of that, is simply a representation of the event. It could be considered a principle provided it facilitates rational interpretation of experimental results overwhelmingly. Thus, experimental results of folding kinetics for numerous proteins are viewed in model perspectives in order to examine the success of a model. The task here is to examine which model best explains the observation of the misfolded protein elucidated in this work.

5.2. Theory-based Models and Protein Misfolding

These models subtly elude the misfolding phenomenon. Although the funnel model held right from the beginning that proteins in the folding run may become trapped in shallow minima at later stages because of certain non-native contacts [19, 20], occurrence of large scale misfolding is not explained. In the case of cytochrome *c* for example, the NCO species (see results) is actually a trapped misfold, because structurally NCO and N are virtually identical except for the heme ligand. The ligand is the extrinsic CO for the former and M80 for the latter. The NCO$\rightarrow$N folding transition whose rate is limited by the thermal dissociation of CO has actually been studied by laser photolysis and interpreted in terms of glassy trapped dynamics as held by the funnel model [21]. However, the large scale misfolding in the refolding of alkaline cyt-CO shown here is difficult to interpret in the funnel perspective.

Quick and large-scale misfolding is difficult to envisage in the theoretical models. Also, these models refrain from using on-pathway and off-pathway terminologies. Nonetheless, an intermediate may be called off-pathway if the structure does not increasingly resemble the native state while on the folding run [2]. For example, numerical results of Monte Carlo simulations of protein-like polymers on a lattice [22-27], and molecualrt dynamics simulations of heteropolymers in continuum [28, 29] show that a heteropolymer/protein quickly acquires a compact structure carrying native as well as non-native contacts. In the course of further evolution, some native-like structural elements are dissolved while others are formed. Experimental results for folding of β-lactoglobulin for which a non-native α-helical conformation formed during the burst folding is transformed to the native β-sheet conformation have been interpreted in this scenario [30]. Similarly, transient formation and disappearance of non-native β-sheet during the folding of α-lactalbumin has been reported [31]. Thus, overwhelming unfolding of a misfolded state appears not considered by theoretical models.

5.3. Classical Models and Protein Misfolding

Being based on the principle of microscopic reversibility and the law of mass action, all of such models allow misfolding of various degrees. In the classical realm, an off-pathway intermediate can be operationally defined as the one that has to undergo, because of its arrival at the dead-end, a large scale unfolding to resemble the initial unfolded state both structurally and energetically. In the present case, a comparison of the chevron gradients for the unfolding and pretransition regions of GdnHCl concentration (Figure 3c) illustrates this point. An on-pathway intermediate, on the other hand, does not have to unfold structurally except for occasional rupture of a few long range interactions or rearrangements of folding chain segments such that it can continue to fold to the native state. Both off- and on-pathway intermediates can result from the initial chain collapse and the formation of local interactions such that both states are lower in energy relative to the unfolded state, and both of them would appear to reduce the conformational search problem.

It should be mentioned though that the phenomenon of the deviation from the predicted linear dependence of logarithm of folding rate constant on denaturant can be both qualitatively interpreted and quantitatively modeled in terms of on-pathway as well as off-pathway placement of an intermediate [32, 33]. Whether intermediates are on the pathway or are merely side precuts is one of the longest-running controversies [34]. While the on-pathway intermediates would appear conducive to folding because they are expected to speed up folding by guiding the polypeptide chain to the native state, the role of off-pathway intermediates becomes less obvious. Here, experimental record of a fundamental kinetic criterion, namely, inversion of the denaturant dependence of the observed folding rate constant reveals the off-pathway dead-end mechanism.

5.4. How Could a Polypeptide Fold to the Dead End?

The arrival of a protein chain at the dead-end is possibly a consequence of unsuccessful search for the correct transition state for folding. This possibility is best illustrated by the topology search model of folding advanced by several laboratories in recent years (for example, [14-16, 35, 36]). As soon as a refolding condition is created, the unfolded chains (U) initially at low-energy equilibrium states are driven instantaneously to new higher energy level(s) [37]. The excited chains, now in a nonequilibrium state, respond to the new solvent condition where only a few, if any, denaturant molecules can affect the nascent polypeptides (U′). This ultrafast relaxation process that may involve collapse or chain contraction, desolvation, and/or redistribution of backbone dihedral φ,ψ angles [16, 35, 38, 39] are missed out in millisecond stopped-flow measurements, and is termed burst folding phase. Because the U→U′ is a submillisecond process and the experimentally observed folding time falls in millisecond to seconds range, it is the U′→N transition that sets the folding time. Further, for apparently two-state fast-folding proteins like ferrocyt *c* [40, 35] and Csp B [41], where post-barrier kinetic intermediates do not accumulate, folding from the transition state to the native state will be rapid and barrier-free. These considerations imply that almost the entire length of the observed folding time is the time needed for the U′ state to find a transition state. It consumes time because of the search-find nature of the process [11, 14, 16, 35]. It is thermodynamically constrained because a stable transition state that can support forward folding is to be found. Search continues until a relatively low-energy transition state stabilized by a few correctly formed native-like tertiary contacts is achieved. This is the essence of the topology search model [14, 16, 35, 42] that draws support from observations that folding rates and mechanisms often appear to be largely determined by the native-state topology or contact order [43, 44]. Native-like chain topology stipulates certain long-range interactions, not local contacts as viewed by theory [37, 45]. In addition, being supported by only a few tertiary contacts the transition state barrier must be less organized and therefore more unfolded-like, consistent with $\alpha<0.3$. Finally, because the topology search is an unsystematic process, the possibility of error commitment is not excluded. A wrong transition state should lead to an off-pathway structure, a convincing experimental evidence of which is provided in this work.

The abortive structure then unfolds, and the latter searches for the correct transition state. When found, folding to the native state is rapid. Achieving the transition state is analogous to the assumption of transition state theory that the systems that have surmounted the col in the direction of products do not turn back. Kinetic intermediates, I_n, of sufficient lifetime, when present, populate the post-barrier pathway.

Finally, inadequacy of data at hand does not allow a structural description of the abortive intermediate. It is also clear that a single-exponential decay of proton occupancy as a function of time (Figure 6) means accumulation of no other structural intermediate. Thus, the refolding of alkaline cyt-CO can be schematized classically as the following:

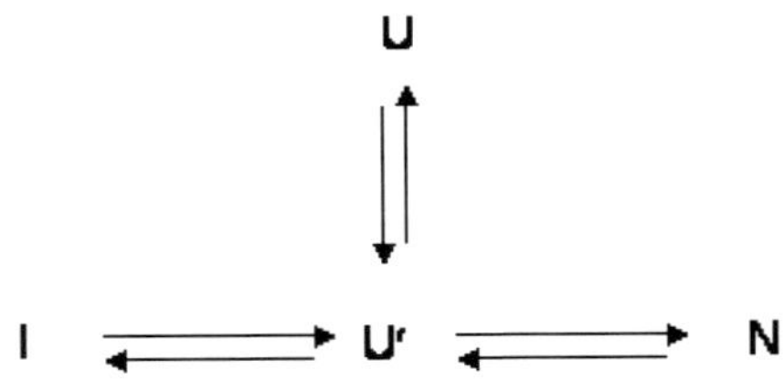

CONCLUSION

The folding time of cytochrome *c* is the time taken for nascent unfolded poly peptide chain to search the correct transition state topology. The search is unsystematic, and in this regime the polypeptide may achieve a wrong transition state leading to a dead-end structure. The polypeptide can nearly or completely misfold in the test tube as some do *in vivo*.

REFERENCES

[1] Onuchic, J. N., Wolynes, P. G., Luthey-Schulten, Z., and Socci, N. D. *Proc. Natl. Acad. Sci. USA* 1995, *92*, 3626-3630.

[2] Dill, K. A., and Chan, H. S. *Nat. Struct. Biol.* 1997, *4*, 10-19.

[3] Onuchic, J. N., Luthey-Schulten, Z., and Wolynes, P. G. *Annu. Rev. Phys. Chem.* 1997, *48*, 545-600.

[4] Socci, N. D., Onuchic, J. N., and Wolynes, P. G. *Proteins* 1998, *32*, 136-158.

[5] Schonbrun, J. & Dill, K. A. *Proc. Natl. Acad. Sci USA* 2003, *100*, 12678-12682.

[6] Onuchic, J. N., and Wolynes, P. G. *Curr. Opin. Struct. Biol.* 2004, *14*, 70-75.

[7] Baldwin, R. L. *J. Biomol. NMR* 1995, *5*, 103-109.

[8] Bryngelson, J. D., Onuchic, J. N., Socci, N. D., and Wolynes, P. G. *Proteins: Struct. Funct. Genet. 1995, 21,* 167-195.

[9] Onuchic, J. N., Socci, N. D., Luthey-Schulten, Z., and Wolynes, P. G. *Fold. Des.* 1996, *1*, 441-450.

[10] Bai, Y., Sosnick, T. R., Mayne, L., and Englander, S. W. *Science* 1995, *269*, 192-197.

[11] Rumbley, J., Hoang, L., Mayne, L., and Englander, S. W. *Proc. Natl. acad. Sci. USA*, 2001, *98*, 105-112.

[12] Hoang, L., Bédard, S., Krishna, M. M. G., Lin, Y., and Englander, S. W. *Proc. Natl. Acad. Sci. USA*, 2002, *99*, 12173-12178.

[13] Krishna, M. M. G., Lin, Y., Rumbley, J. N., and Englander, S. W. *J. Mol. Biol.* 2003, *331*, 29-36.

[14] Makarov, D. E., and Plaxco, K. W. *Protein Sci.* 2003, *12*, 17-26.

[15] Sosnick, T. R., Mayne, L., Hiller, R., and Englander, S. W. *Nat. Struct. Biol.* 1994, *1*, 149-156.

[16] Sosnick, T. R., Mayne, L., and Englander, S. W. *Proteins* 1996, *24*, 413-426.

[17] Santoro, M. M., and Bolen, D. W. *Biochemistry* 1988, *27*, 8063-8068.

[18] Silow, M., and Oliveberg, M. *Biochemistry,* 1997, *36*, 7633-7637.

[19] Bryngelson, J. D., and Wolynes, P. G. *Proc. Natl. Acad. Sci. USA* 1987, *84,* 7524-7528.

[20] Wolynes, P. G., Onuchic, J. N., and Thirumalai, D. *Science* 1995, *267*, 1619-1620.

[21] Yadaiah, M., Kumar, R., and Bhuyan, A. K. *Biochemistry* 2007, *46,* 2545-2551.

[22] Camacho, C. J., and Thirumalai, D. *Proc. Natl. Acad. Sci. USA* 1993, *90,* 6369-6372.

[23] Gutin, A. M., Abkevich, V. I., and Shakhnovich, E. I. *Biochemistry* 1995, *34,* 3066-3076.

[24] Miller, R., Danko, C. A., Fasolka, M. J., Balazs, A. C., Chan, H. S., and Dill, K. A. *J. Chem. Phys.* 1992, *96,* 768-780.

[25] Sali, A., Shakhnovich, E., and Karplus, M. *Nature* 1994, *369,* 248-251.

[26] Sali, A., Shakhnovich, E., and Karplus, M. *J. Mol. Biol.* 1994, *235,* 1614-1636.
[27] Skolnick, J., and Kilinski, A *Science* 1990, *250,* 1121-1125.
[28] Guo, Z., Thirumalai, D., and Honeycutt, J. D. *J. Chm. Phys.* 1992, *97,* 525-535.
[29] Honeycutt, J. D., and Thirumalai, D. *Proc. Natl. Acad. Sci. USA* 1990, *87,* 3526-3529.
[30] Hamada, D., Segawa, S, I., and Goto, Y. *Nat. Struct. Biol.* 1996, *3,* 868-873.
[31] Troullier, A., Reinstadler, D., Dupont, Y., Naumann, D., and Forge, V. *Nat. Struct. Biol.* 2000, *7,* 78-86.
[32] Wildegger, G., and Kiefhaber, T. *J. Mol. Biol.* 1997, *270,* 294-304.
[33] Kumar, R., and Bhuyan, A. K. *Biochemistry* 2005, *44*, 3024-3033.
[34] Udgaonker, J. B., and Baldwin, R. L. *Nature* 1988, *335,* 694-699.
[35] Prabhu, N. P., Kumar, R., and Bhuyan, A. K. *J. Mol. Biol.* 2004, *337*, 195-208.
[36] Bhuyan, A. K., Rao, D. K., and Prabhu, N. P. *Biochemistry* 2005, *44,* 3034-3040.
[37] Shakhnovich, E. I. *Curr. Opin. Struct. Biol.* 1997, *7*, 29-40.
[38] Sosnick, T. R., Mayne, L., and Englander, S. W. *Proteins* 1997, *24*, 413-426.
[39] Jacob, J., Krantz, B., Dothager, R. S., Thiyagarajan, P., and Sosnick, T. R. *J. Mol. Biol.* 2004, *338*, 369-382.
[40] Bhuyan, A. K., and Udgaonkar, J. B. *J. Mol. Biol.* 2001, *312,* 1135-1160.
[41] Schindler, T., Herrler, M., Marahiel, M. A., and Schmid, F. X. *Nature Struct. Biol.* 1995, *2,* 663-673.
[42] Bai, Y., Zhou, H., and Zhou, Y. *Protein Sci.* 2004, *13*, 1173-1181.
[43] Abkevich, A. I., Gutin, A. M., and Shakhnovich, E. I. *J. Mol. Biol.* 1995, *252*, 460-471.
[44] Bhuyan, A. K., and Udgaonkar, J. B. *Proteins* 1998, *32*, 241-247.
[45] Xu, Y., Mayne, L. C., and Englander, S. W. *Nat. Struct. Biol.* 1998, *5*, 774-778.

In: Nuclear Magnetic Resonance (NMR) ISBN: 978-1-63117-768-2
Editor: Doddapuneni Krishna Rao © 2014 Nova Science Publishers, Inc.

Chapter 4

AROMATIC RING-FLIP MOTIONS IN PROTEINS BY EXSY NMR

Doddapuneni Krishna Rao[1,*] and Abani K. Bhuyan[2]

[1]Singapore BioImaging Consertum, A*STAR, Singapore

[2]School of Chemistry, University of Hyderabad, India

1. ABSTRACT

Dynamics of large-amplitude conformational motions in proteins are complex and less understood, although these processes are intimately associated with structure, folding, stability, and function of proteins. A large set of spectra obtained by cross-relaxation suppressed exchange NMR spectroscopy (EXSY) is used to study the 180° flipping motion of the Y97 ring of horse ferricytochrome *c* as a function of near-physiological temperature in the 288-308 K range. With rising temperature, the ring-flip rate constant makes a continuous transition from Arrhenius to anti-Arrhenius behavior through a narrow Arrhenius-like zone. This behavior is seen not only for the native state of the protein, but also for native-like states generated by adding subdenaturing amounts of guanidine deuterochloride (GdnDCl). Moderately destabilizing concentrations of the denaturant (1.5 M GdnDCl) completely removes the Arrhenius-like feature from the temperature window employed. The Arrhenius to anti-Arrhenius transition can be explained by the heat capacity model where temperature strengthens

* E-mail: kri_uoh@yahoo.co.in.

ground state interactions, perhaps hydrophobic in nature. The effect of the denaturant may appear to arise from direct protein-denaturant interactions that are structure-stabilizing under subdenaturing conditions. The temperature distribution of rate constants under different stability conditions also suggests that the prefactor in Arrhenius-like relations is temperature dependent. Although the use of the transition state theory (TST) offers several challenges associated with data interpretation, the present results and a consideration of others published earlier provide evidence for complexity of ring-flip dynamics in proteins.

2. Introduction

In recent years, structural studies by X-ray techniques [1] and NMR spectroscopy [2, 3], and mutational analyses [4, 5] have greatly facilitated relating structure, conformational changes, function, and stability of proteins. Investigations of protein dynamics are still in its infancy though [6]. Internal dynamics that include small amplitude collective motions [7] and large amplitude cooperative breathing modes [8], and other structural fluctuations and deformations at the subglobal level [9] are relatively less understood. The complexity of certain of these processes has been recognized for quite some time in both experimental [6, 10, 11] and theoretical studies [12, 13]. For analytical solutions of experimental data, they often appear complex because of either unusual time evolution or complex temperature dependence or both. In theoretical studies, the complexity surfaces when a suitable reaction coordinate and an appropriate transition state along the coordinate is difficult to define. There, of course, is the problem of current atomistic simulations not reaching times longer than microseconds. However, a comprehensive description of protein motions is important not only for connecting structure and function to dynamics, but also for understanding their role in folding and stability of native and natively unfolded states [14-16].

This chapter presents the temperature and denaturant related complexity of the ring motion of Y97 side chain (180° rotational jump of the ring about the C_β–C_γ bond axis) of horse ferricytochrome *c* observed by cross-relaxation suppressed exchange NMR spectroscopy (EXSY). The rate of the process does not follow Arrhenius temperature dependence under native and native-like conditions. The changes in the rate-temperature trend under different conditions of protein stability appear to indicate temperature dependence of the Arrhenius-type prefactor. Data are analyzed invoking a heat capacity model that assumes a considerable change in the ground-state heat capacity with temperature.

Starting with the seminal work of Wüthrich and Wagner (1975) on BPTI, the rate-temperature relationship for aromatic ring flip has been reported for just about four proteins to date: BPTI [17-20], ferrocytochrome *c* [21], iso-2-cytochrome *c* [22], and HPr protein [23]. Although the activation parameters from all data sets, generated largely by simulation of variable temperature NMR spectra, were extracted by assuming temperature independence of enthalpy and entropy in the Eyring equation, it is believed that they are generally temperature dependent to different extents, and hence could be modeled better by using temperature dependent functions. Indeed, this was indirectly observed in a revisit to the BPTI Y35 ring-flip dynamics [18]. Also, rate-temperature curves calculated for several cases [20] using activation parameters published earlier indicate visibly non-Arrhenius behavior. Together with the observations of the present study of ferricytochrome *c*, ring-flip dynamics appear generally complex.

3. Materials and Methods

All experiments were performed using cytochrome *c* (Type VI from Sigma) and GdnHCl (from USB, GE Healthcare). The labile hydrogens of both of which were preexchanged for deuterium. Exposure of D_2O solutions of cytochrome *c* to pH 10 at 60°C for an hour followed by lyophilization of the neutralized solution caused complete deuteration of the labile proton sites. GdnDCl was prepared by repeated lyophilization of D_2O solutions of the denaturant.

3.1. NMR Spectroscopy

A 2-mM solution of cytochrome *c* containing a desired concentration of GdnDCl was prepared in 50 mM phosphate buffer in D_2O, pH 6.5. The same sample was used to take variable-temperature NOESY and EXSY spectra. Each 2D spectrum was of 400 t_1 increments and 2048 t_2 data points over a spectral width of 18 ppm. Quadrature detection along the indirectly detected dimension was achieved by the States-TPPI method. For exchange rate measurement at a given concentration of GdnDCl and temperature, EXSY spectra [24] were recorded using five mixing times (10, 20, 30, 40, and 50 ms). The details of the mixing pulse scheme are shown in Figure 1. With 128 scans averaged for each t_1, the time required for a 2D spectrum was over 16 hours at 400 MHz. Spectra were recorded using a Bruker spectrometer, and processed and analyzed with XWIN NMR (Bruker) and Felix (Accelrys) softwares.

4. RESULTS AND DISCUSSION

4.1. Y97 Ring-flip Rate Constant As a Function of Protein Stability Coordinate Measured by NMR Spectroscopy

The earliest NMR investigations of protein dynamics that employed techniques of 1D time-resolved saturation transfer, isotope-edited spectroscopy in conjunction with spectra simulation, and analysis of temperature dependence of resonances have made several fundamental contributions to the understanding of aromatic ring rotational motions in proteins [17, 20, 22, 25]. The advent of 2D NMR spectroscopy has greatly facilitated studies of such dynamic processes that are too slow to affect the lineshapes [26, 27]. Subsequent introduction of pulse sequences designed to produce chemical exchange cross-peaks by elimination of cross-relaxation and spin coupling-induced coherence transfer effects in 2D spectra [24, 28] has further augmented the NMR approach to studies of exchange dynamics in proteins. Indeed, numerous recent work have employed EXSY to investigate dynamic effects; particularly, the ring inversion phenomenon in biology and chemistry [20, 29-32].

With the objective of characterizing the NMR-defined slow dynamic processes in cytochrome *c* under native to mild denaturing conditions, It has collected a large set of exchange spectra recorded at several low concentrations of GdnDCl, each as a function of temperature, and in turn as a function of mixing time in the 0-50 ms range. This approach allows mapping the activation parameters for the rate processes, as the protein is progressively destabilized. Figure 1 presents excerpts from the data set showing the extraction and temperature dependence of cross peaks due to rotational motions of Y97 and F10 rings in native cytochrome *c*. Another ten cross peaks assigned to P30 $C_\beta H$, L68 $C_\delta H_3$, Y48/T42 $C_\alpha H$, T28 $C_\gamma H$, F36/T47 $C_\beta H$, K22/G23 $C_\alpha H$, C14/Q16 $C_\gamma H$, $F82_{2,6}$, F46 C_3H, and H26 $C\beta$ appear in the aliphatic region of the native-state spectrum at 22°C.

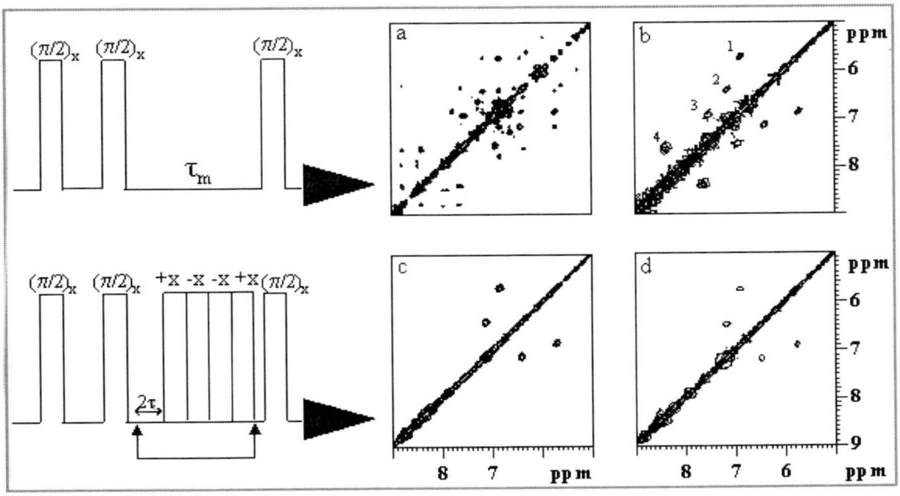

Figure 1. Relaxation suppressed EXSY largely eliminates NOESY peaks. (a) The aromatic region of a NOESY spectrum of ferricyt *c* at 22°C, pH 6.3. The same region of EXSY spectra recorded with the same sample at 20°C (b), 30°C (c), and 35°C (d). Cross peak labels in (b) are 1: $Y97_{3,5}$; 2: $Y97_{2,6}$; 3: $F10_{2,6}$; and 4: $F10_{3,5}$. The pulse sequences shown at the left are for NOESY (top) and EXSY (bottom).

To calculate the rate coefficients and activation parameters of the dynamic processes along the protein stability coordinate, it is necessary that the set of exchange-connected cross peaks in the NMR spectrum consistently appear at all concentrations of the denaturant in a suitable range of temperature. However, since both temperature and denaturant affect the dynamics, those processes the interconversion times of which are shorter or longer than the mixing time of the NMR experiment fail to produce cross peaks; shorter interconversion times lead to exchange broadening, and longer interconversion times make only a minute fraction of exchanging population available for interrogation. Such conditions do not produce quantifiable cross peaks. The requirement of the suitable exchange time severely constrains the number of cross peaks, and hence the dynamic processes that can be examined in detail. Two cross peaks, $Y97_{3,5}$ and $Y97_{2,6}$, (peaks 1 and 2, respectively, in Figure 1b) corresponding to 180° rotational motion of the Y97 ring are consistently observed in the 288-313 K temperature range at 0-1.5 M GdnDCl concentration, and were used for further analysis. The ring-flipping rate constant, k_{flip}, was calculated according to Ernst et al. [26].

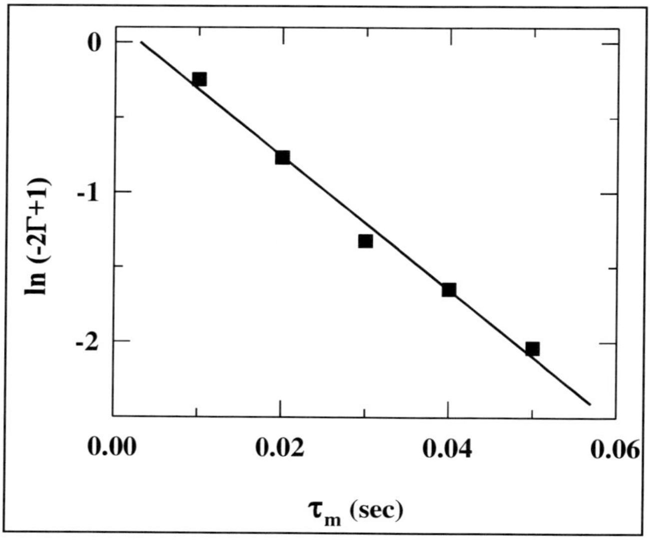

Figure 2. Extraction of the flip rate constant for the Y97 ring in the native protein at 20°C, pH 6.3. The experimental procedure and data analyses are detailed in the text. Results are in table 1.

$$\ln(-2\Gamma + 1) = -2k_{flip}\,\tau_m, \tag{1}$$

where $\Gamma = V_c/(V_c + V_d)$ with V_d and V_c the diagonal and cross-peak volumes, respectively, and τ_m is the mixing time.

Figure 2 shows an example of the plot of ln(-2Γ+1) vs τ_m.

Values of k_{flip} thus obtained should preferably be corrected for the viscosity effect due to GdnDCl. Although the relative viscosity of the medium increases only marginally in the range of denaturant concentrations used in this study (0-1.5 M), a correction can be introduced according to $k_{flip} \propto 1/\eta$, where η is the bulk viscosity. To eliminate the effect of viscosity due to GdnDCl on the ring-flip rate constant, it has simply used relative viscosity η/η_o, where η_o is the viscosity in the absence of GdnDCl. The corrected k_{flip} is then given by the relation

$$\ln k_{flip}^{corr} = \ln k_{flip} + \ln\left(\frac{\eta}{\eta_o}\right). \tag{2}$$

To obtain the values of η/η_o, viscosity values of aqueous solutions of GdnDCl from literature [33] were used to find the following empirical relation

$$\frac{\eta}{\eta_o} = 1 + 0.005[D]^{2.6144} + 0.018[D]^{0.6594} + 0.01213[D]^{0.6636}, \tag{3}$$

where [D] represents molar concentration of GdnDCl.

4.2. Non-Arrhenius Behavior for the Ring-flip Rate Constant

The temperature dependence of k_{flip} for Y97 at four concentrations of GdnDCl is presented in Figure 3. The plots of $\ln(k_{flip}/T)$ against 1000/T are not at all linear. The gradients for the curves at 0, 0.5, and 1 M GdnDCl make continuous transition from negative at low temperatures to positive at high temperatures, more accentuated for the latter two concentrations of the denaturant. In general, it has identified the curved temperature dependence of rate constants with non-Arrhenius behavior, which could be operationally termed Arrhenius-like (negative slope) or anti-Arrhenius (positive slope) for a small segment of the temperature axis. Indeed, one notices a gradual turn from Arrhenius-like to anti-Arrhenius behavior as the temperature is raised from 15 to 40 °C. At 1.5 M GdnDCl, where the protein is still native-like but denaturing conditions are approached [34], the flip rate constant shows only anti-Arrhenius temperature dependence.

To explain the diminishing enthalpic but growing entropic contribution to the ring isomerization barrier with increasing temperature, it is chose to use the transition state theory (TST) rate expression. For justification of this choice, a detailed consideration of reaction rate theory more appropriate for protein reactions is required, a general understanding of which is still not available. It is also uncertain if the ring motion is fully in the diffusive regime, so a Kramers-like rate description becomes essential. The magnitude of the internal friction, and the extent of influence of system-solvent collisions on the ring-flip reactive trajectories across the barrier are other uncertainties. Further, the ring motion is unlikely to be associated with a low and flat barrier for which one might expect a diffusive rate. On the other hand, within the limitations of presently available reaction rate theories and microscopic models, a full justification for the adequacy of TST is difficult to achieve.

Table 1. Values of k_{flip} (s^{-1}) as a function of temperature and GdnDCl concentration

Temperature (K)	GdnDCl (M)			
	0	0.5	1.0	1.5
289	4.60	7.56	9.47	10.66
293	13.25	11.85	10.41	9.39
298	22.30	14.56	15.75	2.10
303	33.40	13.91	3.36	0.89
306	26.48	–	–	–
308	–	3.38	–	–
313	–	0.68	–	–

Error in determination of rates: 2-20%.

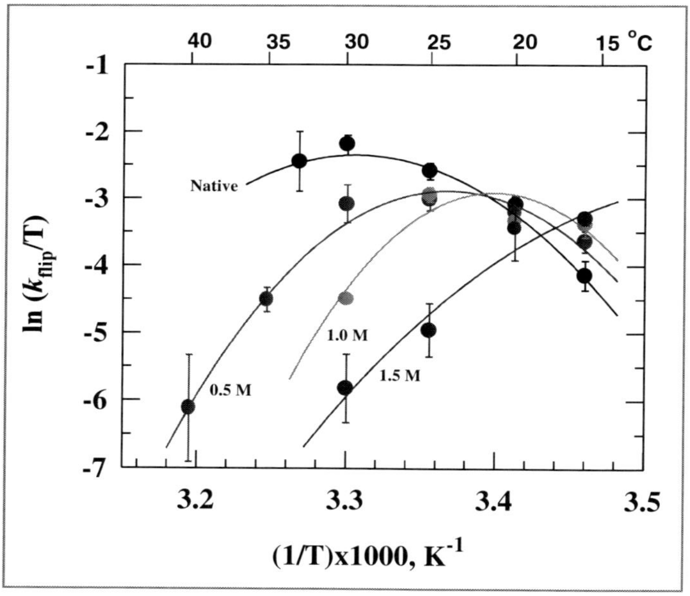

Figure 3. Temperature dependence of the Y97 ring-flip rate constant at four concentrations of GdnDCl: 0 M (black), 0.5 M (red), 1.0 M (yellow), and 1.5 M (blue). Solid lines are fits to data by equation 6. Values of $\Delta H^{\ddagger}$, $\Delta C_p^{\ddagger}$, and $\Delta S^{\ddagger}$ obtained from the fits are listed in Table 2.

Table 2. Activation parameter for flipping motion of the Y97 ring

GdnDCl/M	Thermodynamic Parameters		
	$\Delta H^{\ddagger}$ (kcal mol^{-1})	$\Delta C_p^{\ddagger}$ (kcal mol^{-1} 1000K^{-1})	$\Delta S^{\ddagger}$ (kcal mol^{-1} 1000K^{-1})
0.0	15.3	-3.448	0.019
0.5	-4.9	-4.697	-0.049
1.0	-25	-6.645	-0.117
1.5	-38	-2.196	-0.165

Estimated error range for these values is 3-5%.

To make progress with the data at hand, the analysis was continued with the Eyring-type expression

$$\ln k_{\text{flip}} = \ln A - \frac{\Delta G^{\ddagger}}{RT}, \tag{4}$$

where $\Delta G^{\ddagger} = \Delta H^{\ddagger} - T\Delta S^{\ddagger}$ is the barrier free-energy. The exact nature of the prefactor A is not known; it is perhaps determined by microscopic dynamics entailed in barrier crossing [14]. $\Delta G^{\ddagger}$ is expanded by casting the enthalpy and entropy changes in their basic forms

$$\Delta H_T^{\ddagger} = \Delta H_{T_0}^{\ddagger} + \int_{T_0}^{T} \Delta C_P{}^{\ddagger} dT, \quad \text{and} \quad \Delta S_T^{\ddagger} = \Delta S_{T_0}^{\ddagger} + \int_{T_0}^{T} \Delta C_P{}^{\ddagger} \frac{dT}{T}, \tag{5}$$

which allow evaluation of enthalpy and entropy changes at temperatures T with respect to a reference temperature T_0. $\Delta C_p{}^{\ddagger}$ is the difference in heat capacity of the transition state and the initial state.

T_0 is set at 25°C (see below), although for accurate calculations of heat-induced conformational transitions, the midpoint temperature should be used. Thus, the equation

$$\ln k_{\text{flip}} = \ln A - \frac{1}{RT}\left[\Delta H_{T_0}^{\ddagger} + \Delta Cp^{\ddagger}(T - T_0) - T\left(\Delta S_{T_0}^{\ddagger} + \Delta C_P{}^{\ddagger} \ln\left(\frac{T}{T_0}\right)\right)\right] \tag{6}$$

was used to fit the data shown in Figure 3. Values of $\Delta H^{\ddagger}$, $\Delta S^{\ddagger}$, and $\Delta C_p^{\ddagger}$ for all four curves are listed in Table 2.

4.3. Variation of the Activation Parameters Along the Protein Stability Coordinate

The details of the ring-flip activation parameters in the presence of low concentrations of GdnDCl could potentially provide important information regarding collective motion, stability, and interactions of the relevant structural elements under subdenaturing to denaturing conditions. For scaling the activation parameter, it is reasonable to assume that the changes of the differences in thermodynamic properties between the transition state and the ground state under different conditions of temperature and stability are largely due to effects at the ground state level [35], because the state variables are expected to cause relatively less changes in the thermodynamic properties of a changing transition state which is structurally diminished and distorted.

The temperature gradient of $\Delta H^{\ddagger}$ (Figure 4a) is negative for all concentrations of GdnDCl. The curves precisely indicate the turn from Arrhenius to anti-Arrhenius through a relatively smaller Arrhenius-like zone of temperature. For all conditions, the value of $\Delta H^{\ddagger}$ passes from positive to negative with increasing temperature registering 'zero' at 298 K, a consequence of choosing T_o=298 K. Obviously, T_o cannot be 298 K all along the stability coordinate. Even for the native state, the chosen T_o may be way off. It is hard to determine the melting temperature for the specific subglobal part of the protein that is involved in the ring isomerization process. Nonetheless, the arbitrary choice of T_o can still provide qualitative information and relative values of activation parameters, since the temperature gradient of $\Delta H^{\ddagger}$ will depend little on T_o.

The increase in the slope of the enthalpy curves with increments of GdnDCl up to 1 M (Figure 4a) suggests that in the subdenaturing limit of its concentration, the denaturant acts to increase the difference in heat capacity between the transition state and the ground state ($\Delta C_p^{\ddagger}$). At 1.5 M GdnDCl, denaturing conditions are approached, and $\Delta C_p^{\ddagger}$ begins to shrink. This is also seen from the values listed in Table 2, where the –ve sign for $\Delta C_p^{\ddagger}$ indicates larger values for the ground state C_p. An interpretation of this result could be that within the subdenaturing limit of its concentration, GdnDCl somehow modulates the solvent properties and introduces additional interactions by virtue of its mechanism of action on proteins. This aspect of GdnDCl-protein interaction and the consequences thereof is being discussed in detail in a later section.

Like the temperature dependence of $\Delta H^{\ddagger}$, the value of $\Delta S^{\ddagger}$ passes from positive to negative with increasing temperature for all concentrations of GdnDCl (Figure 4b). Clearly, the entropic contribution to the barrier energy dominates in the respective anti-Arrhenius temperature zones (Figure 3). The slope of the entropy-temperature curve increases in going from 0 to 1 M denaturant, suggesting that in the presence of increasing subdenaturing concentrations of GdnDCl, relatively large-scale conformational constraint must be negotiated to make to the transition state. The constraint is generated by the binding interaction of the denaturant with the protein. The decrease of the slope at 1.5 M denaturant is due to the onset of denaturation, where entropy-lowering restraints in the ground state begin to weaken because of increasing structure-breaking action of the denaturant in at least that part of the protein molecule which harbors the flipping ring.

Temperature variation of $\Delta G^{\ddagger}$ for different concentrations of GdnDCl, plotted in figure 4c, shows vertical shift of energy and horizontal shift of the temperature corresponding to the energy minimum. The magnitude and the sign of these shifts are specific to the reference temperature chosen (T_o=298 K). For the same reason, $\Delta G^{\ddagger}$ value for the native-state protein comes out as negative, and therefore, the shifts obtained from Figure 4c must be scaled relatively. Figure 4d shows how the temperature corresponding to minimal $\Delta G^{\ddagger}$ decreases with increasing amount of GdnDCl; this is due to the general destabilizing action of the denaturant. On the other hand, the increase in the minimal $\Delta G^{\ddagger}$ with denaturant (Figure 4e), implying a lowering of the free energy of the ground state relative to the transition state, is due to the stabilizing action of the denaturant. The increase is nonlinear though, suggesting that the initial protein stabilization caused by subdenaturing amounts of GdnDCl withers as denaturation begins to set in (see below). By magnitude, the increase of $\Delta G^{\ddagger}$ as a function of GdnDCl is nearly the same within the temperature range employed in our experiment (marked by dotted line in Figure 4c). The observed variation of rates in this temperature range, however, appears complex. For example, the k_{flip} value for 0 M GdnDCl relative to that for 1 M GdnDCl is an order of magnitude larger at 30°C, is matching at ~20°C, and is smaller at lower temperature. These differences are not due to error of measurement.

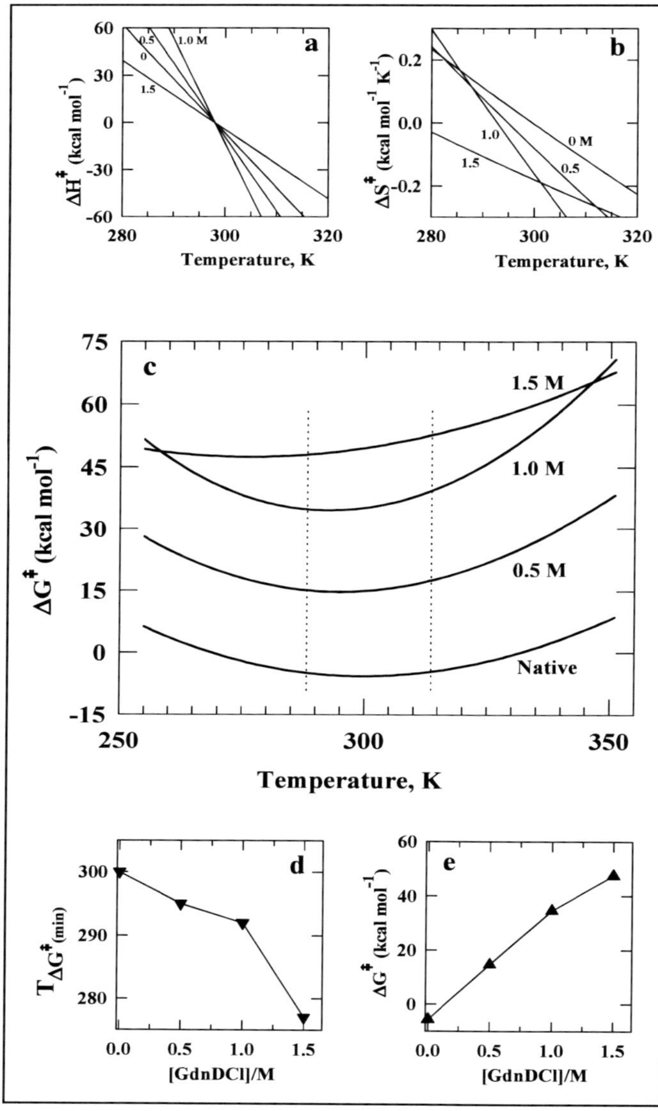

Figure 4. Thermodynamic parameters for the Y97 ring-flip dynamics calculated from the data shown in Figure 3. Temperature distribution of (a) $\Delta H^{\ddagger}$, (b) $\Delta S^{\ddagger}$, and (c) $\Delta G^{\ddagger}$ for the four concentrations of GdnDCl indicated. The vertical dotted lines in panel (c) show the temperature range used in this study. (d) GdnDCl distribution of the temperature corresponding to minimal $\Delta G^{\ddagger}$. (e) GdnDCl distribution of the minimal $\Delta G^{\ddagger}$.

They rather originate from a temperature dependence of the prefactor *A*, possibly compounded by the effect of GdnDCl on the internal friction of the protein.

4.4. Temperature Dependence of $\Delta G^{\ddagger}$

In Arrhenius-type rate expressions, $\Delta G^{\ddagger}$ should be virtually independent of temperature. However, variable degrees of temperature dependence of $\Delta G^{\ddagger}$ have been described for different systems, including simple metathesis reactions [36, 37] enzyme mechanisms [38], DNA fluctuations [39], protein-ligand reactions [10], and protein refolding kinetics in general [35, 40]. The unusual temperature dependence of certain gas-phase reactions has been explained by a modified transition state theory [37]. For protein reactions, folding processes in particular, two explanations are provided to account for temperature dependence of $\Delta G^{\ddagger}$. The first, explicable by statistical mechanical principles, is derived for a random walk of an activation within a Gaussian distribution of conformations [41], and is often referred to as super-Arrhenius temperature dependence where $exp{-}(\Delta G^{\ddagger}/RT)^2$ is used instead of $exp{-}(\Delta G^{\ddagger}/RT)$ in the rate-temperature expression [42, 43]. The second, based on foolproof experimental evidence for strong temperature dependence of hydrophobic interactions [44, 45], proposes that temperature dependence of $\Delta G^{\ddagger}$ is due to the disruption of large buried apolar surfaces in going from the ground state to the transition state; the associated heat capacity change $\Delta C_p^{\ddagger}$, and hence $\Delta G^{\ddagger}$, is strongly temperature dependent [35, 40]. Another explanation, more appropriate for bimolecular protein reactions, accords complex temperature dependence of $\Delta G^{\ddagger}$ when the binding rate competes with the rates of conformational fluctuations or relaxations of the target site. A well-documented example is rebinding of CO after flash photolysis of carbonmonoxymyoglobin at low temperature [10, 46].

Of these, the second explanation corresponds to the heat capacity model used here for analyzing the ring-flip data. The heat capacity model draws support from strong connections recognized for large amplitude breathing motions of the protein and the ring isomerization event [17, 18, 20, 23, 47]. Since deformations of structure and interactions are involved in large amplitude breathing modes [8, 48], a difference in heat capacity between the transition state and the ground state ($\Delta C_p^{\ddagger}$) is expected. It is also generally accepted that $\Delta C_p^{\ddagger}$ contributes at least in part to non-Arrhenius temperature dependence of protein reactions. The less preference for the use of a super Arrhenius-type model for ring-flip analysis rests on two considerations. First, unlike the folding of protein chains, ring-flip dynamics is not quite a diffusive process in configurational space. Second, the super-Arrhenius relation applies better for an extended temperature range. At temperatures well above the glass transition, the rate coefficient is smaller than the relation predicts [46].

4.5. Challenges Associated with Data Interpretation

The analyses and discussions presented above also illustrate some of the major difficulties associated with data interpretation and calculation of heat capacity differences between ground and excited states. Fundamentally, the use of a kinetic theory and an Eyring-like expression modeled on simple chemical reaction may not necessarily lead to uncontestable insight. Even when granted, the use of an arbitrarily chosen constant reference temperature (T_0=298 K) provides at best a qualitative picture, because the value of T_0 must change according to the protein stability influenced by the denaturant. Accurate values for T_0 relevant for the subglobal part harboring the ring are often difficult to obtain. Another concern is that the ground and excited states vary greatly as a function of denaturant concentration and temperature. Clearly, a detailed knowledge of the landscape topography under denaturing to native-like conditions (i.e., near the bottom of the funnel) is essential for a more quantitative description of the problem. It should mention that the strong temperature dependence of exchange rate of a buried water molecule in BPTI has been previously associated with internal motions involving multiple conformational substate interconversions in a rugged energy landscape [49].

4.6. Structural Basis for Non-Arrhenius Ring-Flip Dynamics in Cytochrome *c*

As mentioned already, fairly good evidence exists for the concerted control of large-scale protein motions on the ring isomerization process. Assuming generality of this, numerous influential work on structure and folding of cytochrome *c* facilitates surmising the motional mode that could drive the Y97 ring-flip process. Y97 is a resident of the C-terminal helix, and it is near this residue that the N- and C-terminal helices form a tight contact to produce an orthogonal geometry of helix docking (Figure 5; see also ref. [50, 51]).

This contact region is made of predominantly hydrophobic residues that are highly conserved, and the majority of the very slowly exchanging amide protons, including residues 94-99, are found on the N-and C-terminal helices [49]. Sufficient exposure of the hydrophobic surface may be required to allow flipping motion of the Y97 ring.

It does not have a direct evidence to show how the motion of the helices arranges for the ring-flip. Reports that an early kinetic intermediate of cytochrome *c* comprises the structures of these two helices [50, 52], the corresponding peptides in solution self-associate [53], and that the associated structure of these helices in the native protein forms a low-energy cooperative unfolding unit simulating the global unfolding behavior [54], all provide a seeming basis to assume that intrinsic association and dissociation or relative sliding motions of the docked helices render Y97 ring flipping.

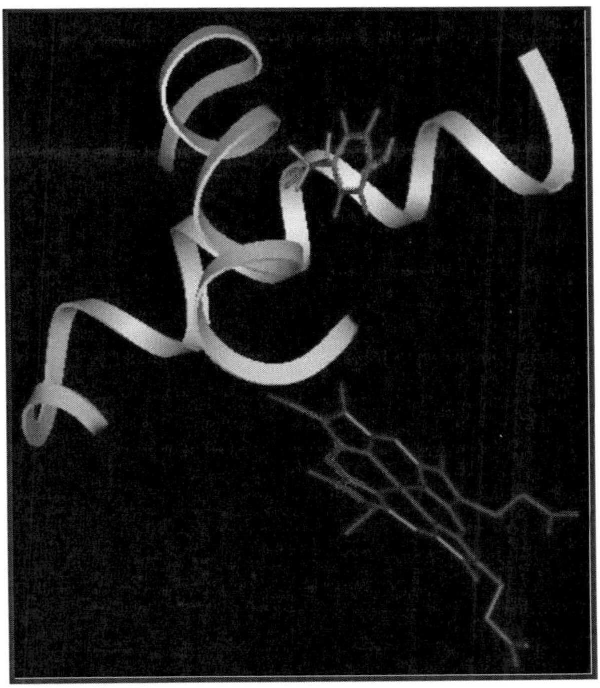

Figure 5. The N- and C-terminal helices of cytochrome *c* with the disposition of the Y97 ring as seen in the crystal structure (PDB file: 1HRC). Also shown is the heme ring.

In thermodynamic terms, the relative motions of the helices that expose the hydrophobic surface in the vicinity of Y97 involve a sizable heat capacity C_p, the magnitude of which increases with strengthening of the hydrophobic interactions by temperature. This motional mechanism must be treated as a possibility. The $\Delta C_p^{\ddagger}$ value is negative for all curves (Table 2), implying that an extensive hydrophobic surface characterizes the ground state, and a large heat is involved in the dissolution of the surface.

4.7. Complex Action of GdnDCl and Temperature on Protein Stability and Ring-flip Rate

One of the major objectives of this study was to learn how the rates of slow dynamic processes, Y97 ring-flip in the context, change as the protein is placed in incrementally destabilizing conditions. A gradual decrease of k_{flip} occurs with increments of GdnDCl, clearly noticeable at temperatures higher than ~20°C. This general decrease can be explained by constrained dynamics of the protein as a result of its interaction with the denaturant. An explanation and the illustration of this effect in the case of cytochrome *c* have been provided from our lab with different set of experiments [55, 56]. Briefly, GdnDCl can interact directly with protein backbone and side-chains by variable-length hydrogen bonding and van der Waals interactions [57-60]. Such interactions produce cross-links or non-specific network of intraprotein interactions [57], leading to protein stiffening, reduction of motional freedom, increased internal friction, and thus entropic stabilization. In the present case also, GdnDCl-induced constraints on the motions of N- and C-terminal helices of cytochrome *c* is likely to retard the ring-flip motion. The protein stabilizing effect of subdenaturing amount of GdnDCl is overrun by its structure unfolding action when used at higher concentrations. The present results indicate that the denaturing effect begins to appear at ~1.5 M GdnDCl.

But the observed k_{flip} dependence on GdnDCl is more complex when the temperature variable is brought into picture. The k_{flip}-temperature curves appear to meet near 20°C, and diverge on either side showing a reversal of temperature dependence (Figure 3), suggesting that the ring flip rate constant near this temperature does not change with GdnDCl as long as the latter is present in subdenaturing concentrations. The slight difference in the behavior of the curve for 1.5 M GdnDCl is likely due to approach of denaturing conditions. This behavior of k_{flip} as a function of denaturant and temperature (Figure 3) is inconsistent with the temperature distribution of $\Delta G^{\ddagger}$ at the four

GdnDCl concentrations (Figure 4c). The vertical dotted lines in Figure 4c show the temperature range used in our experiments, and within this range the variation of $\Delta G^{\ddagger}$ with denaturant is very similar for all temperatures, suggesting that the k_{flip} curves for different denaturant concentrations should not converge at any temperature within this range. This observation should mean that the prefactor *A* is acting in a temperature dependent manner leading to the coalescence of k_{flip} values at ~20°C. But why this happens near 20°C is not clear at present. In Eyring-type relations, *A* generally shows only weak temperature dependence. However, the fits of the data to the heat capacity model with three floating parameters ($\Delta H^{\ddagger}$, $\Delta C_p^{\ddagger}$, $\Delta S^{\ddagger}$) cannot rule out a significant nonlinear dependence of *A* on temperature in a denaturant dependent manner. Studies on the effect of GdnDCl on *A* in Eyring-type expressions will shed light on this issue.

4.8. Aromatic Ring-Flip Motions in Other Proteins

Following the earliest study with BPTI [25], ring-flip dynamics of only four proteins, namely, ferrocytochrome *c* [21], fd bacteriophase [60], yeast iso-2-cytochrome *c* [22], and *S. carnosus* HPr [23] have been studied in some detail, although the problem in BPTI has been revisited on several occasions [17, 18, 20, 47]. In general, the rate strongly varies with packing and dynamics of the ring environment. This is the likely reason why the rates for different aromatic rings within a protein, and for one protein from another could differ, often very significantly, under identical experimental conditions. For example, the rates for Y35 and Y23 of BPTI near room temperature may vary by an order of magnitude, from ~10^2 to ~10^3 s^{-1} [17], and the aromatic rings of fd bacteriophage flip at ~10^6 s^{-1} [61].

Even more interesting is the complex temperature dependence, which the present study exposes in some detail. A review of earlier rate-temperature data for ring dynamics in different proteins also reveals some deviation from simple Arrhenius behavior. The temperature dependence of Y35 flip rate constant in BPTI [17], for example, would appear nonlinear unless large errors in data are allowed. As Karplus points out [13], in the reinvestigation of Y35 flip rate by the use of high resolution EXSY, Otting et al. [18] tie the activation parameters for the flipping process with those for disulfide conformational exchange motion at high temperature. The flip rate at low temperature (<20°C) was not measured. However, since the temperature graph for the activation energy of the disulfide exchange motion shows different dependences at low and high temperatures registering a sharp inflection centered around 20°C, and because the conformational motion and flip motion are linked, the latter should also follow the trend for the former process and hence a nonlinear function of temperature. In a more recent paper, Skalicky et al. [20] used activation parameters published earlier to calculate rate-temperature curves for several residues of BPTI, iso-2-cytochrome *c*, and horse cytochrome *c*. The curves in their semilogarithmic plot of k_{flip} versus T, calculated using $1/k_{flip}=h/(k_BT)\exp[(\Delta H^{\ddagger}/RT-\Delta S^{\ddagger}/R)]$, are clearly not linear. The flip dynamics of Y6 in *S. carnosus* HPr deduced from spectra simulation [23] has been analyzed assuming temperature independence of the activation parameters. Here also, the rate-temperature behavior, especially at 200 Mpa, may not simply be linear; an even closer look is perhaps worth considering. Thus, evidences are rather strong that aromatic ring dynamics in proteins are complex, and they do not follow an Arrhenius-like equation.

Conclusion

Unusual temperature dependence for dynamic processes like ring rotation within a temperature range where spectroscopic measurements do not reveal any change in the time-averaged native structure of the protein is a complex phenomenon by itself. The complexity is compounded when subdenaturing amounts of denaturant act on the protein. Aromatic ring isomerization in dense systems like proteins and polymers involves large amplitude cooperative motion of the ring and certain structural elements, often engaging the chain backbone [18, 62]. It has used the heat capacity model to explain the cytochrome *c* results. Further studies may reveal alternative, perhaps even better models for such phenomena. Irrespective of the merit of the model, the data and evidences presented are compelling that aromatic ring motions are complex.

References

[1] Pellicena, P., Kuriyan, J. *Curr. Opin. Struct. Biol.* 2006, *16,* 702-709.

[2] Grishaev, A., Wu, J., Trewhella, J., Bax, A. *J. Am. Chem. Soc.* 2005, *127,* 16621-16628.

[3] Stangler, T., Hartmann, R., Willbold, D., Koenig, B. W. *Z. Phys. Chem.* 2006, *220,*567-613.
[4] Dill, K. A., Bromberg, S., Yue, K. Z., Fiebig, K. M., Yee, D. P., Thomas, P. D., and Chan, H. S. *Protein Sci.* 1995, 4, 561-602.
[5] Shaw, B. F., Valentine, J. S. *Trends Biochem. Sci* 2007, *32,* 78-85.
[6] Fenimore, P. W., Frauenfelder, H., McMahon, B. H., Young, R. D. *Proc. Natl. Acad. Sci. USA* 2004, *101,* 14408-14413.
[7] Karplus, M. *Methods Enzymol.* 1986, *131*,283-307.
[8] Englander, S. W., Downer, N. W., Teitelbaum, H. *Annu. Rev. Biochem.* 1972, *41*, 903-924.
[9] Ansari, A., Berendzen, J., Bowne, S. F., frauenfelder, H., Iben, I. E. T., Sauke, T. B., Shyamsunder, E., young, R. D. *Proc. Natl. Acad. Sci. USA* 1985, *82*, 5000-5004.
[10] Austin, R. H., Beeson, K. W., Eisenstein, L. *Biochemistry* 1975, *14,* 5355-5373.
[11] Steinbach, P. J., Ansari, A., Berendzen, J. et al., *Biochemistry* 1991, *30,* 3988-4001.
[12] McCammon, J. A., Wolynes, P. G., Karplus, M. *Biochemistry* 1979, *18*, 927-942.
[13] Karplus, M. *J. Phys. Chem. B* 2000, *104,* 11-27.
[14] Portman, J. J., Takada, S., Wolynes, P. G. *J. Chem. Phys.* 2001, *114*,:5082-5096.
[15] Okazaki, K-i., Koga, N., Takada, S., Onuchic, J. N., et al., *Proc. Natl. Acad. Sci. USA* 2006, *103,* 11844-11849.
[16] Doan-Nguyen, V., Loria, J. P. *Protein Sci.* 2007, *16,* 20-29.
[17] Wagner, G., DeMarco, A., Wüthrich, K. *Biohys. Struct. Mechanism* 1976, *2,* 139-158.
[18] Otting, G., Liepinsh, E., Wüthrich, K. *Biochemistry* 1993, *32*, 3571-3582.
[19] Li, H., Yamada, H., Akasaka, K. *Biochemistry* 1998, *37,* 1167-1173.
[20] Skalicky, J. J., Mills, J. L., Sharma, S., & Szyperski, T. *J. Am. Chem. Soc.* 2001, *123*, 388-397.
[21] Campbell I. D, Dobson C. M, Moore G. R, Perkins S. J, Williams R. J. P *FEBS Lett.* 1976, *70,* 96-100..
[22] Nall, B. T., Zuniga, E. H. *Biochemistry* 1990, *29,* 7576-7584.
[23] Hattori, M., Li, H., Yamada, H., Akasaka, K., Hengstenberg, W., Gronwald, W., and Kalbitzer, H. R. *Protein Science* 2004, *13,* 3104-3114.

[24] Fejzo, J., Zolnai, Z., Macura, S., Markley, J. L. *J. Magn. Reson.* 1990, *88*, 93-110.

[25] Wüthrich, K., Wagner, G. *FEBS Lett.* 1975, *50,* 265-268.

[26] Ernst, R. R., Bodenhausen, G., & Wokaun, A. *Principles of Nuclear Nagnetic Resonance in One and Two Dimensions* 1988 (Clarendon Press, Oxford).

[27] Roder, H. *Methods Enzymol.* 1989, *176,* 446-473.

[28] Fejzo, J., Westler, W. M., Macura, S., Markley, J. L. *J. Magn. Reson.* 1991, *92,* 20-29.

[29] Shokhirev, N. V., Shokhireva, T. K., Polam, J. R., Watson, C. T. et al. *J. Phys. Chem. A* 1997, *101,* 2778-2786.

[30] Whittaker, S. B-M., Boetzel, R., MacDonald, C., Lian L-Y et al. *J. Biomol. NMR* 1998, *12*, 145-159.

[31] Lam, P. C-H., Carlier, P. R. *J. Org. Chem.* 2005, *70,* 1530-1538.

[32] Gallego, J. *Nucleic Acid Res.* 2004, *32,* 3607-3614.

[33] Kawahara, K., and Tanford, C. *J. Biol. Chem.* 1966, *241,* 3228-3232.

[34] Kumar, R., Bhuyan, A. K. *Biochemistry* 2005, *44,* 3024-3033.

[35] Oliveberg, M., Tan, Y-J., Fersht, A. R. *Proc. Natl. Acad. Sci. USA* 1995, *92,* 8926-8929.

[36] Benson, S. W., Dobis, O. *J. Phys. Chem. A* 1998, *102,* 5175-5181.

[37] Krasnoperov, L. N., Peng, J., Marshall, P. *J. Phys. Chem. A* 2006, *110,* 3110-3120.

[38] Truhlar, D. G., Kohen, A. *Proc. Natl. Acad. Sci. USA* 2001, *98,* 848-851.

[39] Wallace, M. I., Ying, L., Balasubramanian, S., Klenerman, D. *Proc. Natl. Acad. Sci. USA* 2001, *98,* 5584-5589.

[40] Chan, B-l., Baase, W. A., Schellman, J. A. *Biochemistry* 1989, *28,* 691-699.

[41] Bryngelson, J. D., Wolynes P. G. *J. Phys. Chem.* 1989, *93*:6902-69151995

[42] Bryngelson, J. D., Onuchic, J. N., Socci, N. D., Wolynes, P. G. *Proteins* 1995, *21,* 167-195.

[43] Scalley, M. L., Baker, D. *Proc. Natl. Acad. Sci. USA* 1997, *94,*10636-10640.

[44] Baldwin, R. L. *Proc. Natl. Acad. Sci. USA* 1986, *83,* 8069-8072.

[45] Dill, K. A. *Biochemistry* 1990, *29,* 7133-7155.

[46] Steinbach PJ, Ansari A, Berendzen J et al. *Biochemistry* 1991,*30,*3988-4001.

[47] Wagner, G., Wüthrich, K. *Nature* 1978, *275,* 247-248.

[48] Englander SW, Mayne L. *Annu. Rev. Biophys. Biomol. Struct.* 1992, *21*:243-265.

[49] Denisov, V. P., Peters, J., Hörlein, H. D., Halle, B. *Nat. Struct. Biol.* 1996, *3,* 505-509.

[50] Roder. H., Elöve G. A., Englander S. W. *Nature* 1988, *335*, 700-704.

[51] Bushnell, G. W., Louie, G. V., Brayer, G. D. *J. Mol. Biol.* 1990,*214,* 585-595.

[52] Elöve, G. A., Bhuyan, A. K, Roder, H. *Biochemistry* 1994, *33,* 6925-6935.

[53] Wu, L. C., Laub, P. B., Elöve, G. A., Carey, J. et al., *Biochemistry* 1993, *32,* 10271-19276.

[54] Maity, H., Rumbley, J. N., Englander, S. W. *Proteins* 2006, *63,* 349-355.

[55] Bhuyan, A. K. *Biochemistry* 2002, 41:13386-13394.

[56] Kumar, R., Prabhu, N. P., Yadaiah, M., Bhuyan, A. K. *Biophys. J.* 2004, *87,* 2656-2662.

[57] Dunbar, J., Yennawar, H. P., Banerjee, S., Luo, J., et al. *Protein Sci.* 1997, *6,* 1272-1733.

[58] Makhatadze, G. I., Privalov, P. L. *J. Mol. Biol.* 1992, *226,* 491-505.

[59] Pike, A. C. W., Acharya, R. *Protein Sci.* 1994, *3,* 706-710.

[60] Zarrine-Afsar, A., Mittermaier, A., Kay, L. E., Davidson, A. R. *Protein Sci.* 2006, *15,*162-170.

[61] Gall, C. M., Cross, T. A., DiVerdi, J. A., Opella, S. J. *Proc. Natl. Acad. Sci. USA* 1982, *7,* 101-105.

[62] Khare R, Paulaitis M. E. *Macromolecules* 1995, *28,* 4495-4504.

In: Nuclear Magnetic Resonance (NMR) ISBN: 978-1-63117-768-2
Editor: Doddapuneni Krishna Rao © 2014 Nova Science Publishers, Inc.

Chapter 5

NMR SPECTROSCOPIC CHARACTERIZATION OF DEATH DOMAIN SUPERFAMILY PROTEINS: STRUCTURES, DYNAMICS AND INTERACTIONS

Yufeng Wei[1,*] and Edward C. Twomey[2]

[1]Institute of NeuroImmune Pharmacology, Seton Hall University, South Orange, NJ, US

[2]Integrated Program in Cellular, Molecular and Biomedical Studies, Columbia University, New York, NY, US

ABSTRACT

Death domain superfamily proteins, consisting of death domain (DD), death effector domain (DED), caspase activation and recruitment domain (CARD), and pyrin domain (PYD) subfamilies, are predominantly involved in assembly and activation of apoptotic complexes in the tumor necrosis factor (TNF) receptor-mediated programmed cell death by promoting homotypic protein-protein interactions. Despite low sequence homology across subfamilies, these structural domains are similarly folded into a characteristic antiparallel six-helix bundle structure, and are distinguished by domain-domain interactions and surface features. Death motif superfamily domains possess no catalytic activities, and all their biological functions are

* E-mail: Yufeng.Wei@shu.edu; Fax: +1-973-761-9772; Phone: +1-973-275-2335.

exerted through stimulating specific protein-protein interactions within the subfamily and across a wide variety of biological pathways. Structures of member proteins in all four subfamilies have been delineated by NMR spectroscopy and/or X-ray crystallography. Additionally, biologically and functionally relevant protein dynamics and protein-protein interactions using novel NMR techniques have been the focus of most recent studies. In this chapter, we will review recent advances in the NMR characterization of death domain superfamily proteins in the context of structures, dynamics, and interactions, and the implications for these studies in understanding the cellular processes of apoptosis, inflammation, proliferation, and glucose metabolism.

1. Introduction

Death structural domains are involved in assembly and activation of apoptotic complexes. Death domain superfamily consists of death domain (DD), death effector domain (DED), caspase activation and recruitment domain (CARD), and pyrin domain (PYD) subfamilies. Although sequence homology is low (10-20%) across subfamilies, these domains fold similarly into a signature antiparallel six-helix bundle structure with Greek key topology [1].

These small structural domains (80-90 residues) have no catalytic activities, and all their functions are exerted through invoking protein-protein interactions. The subfamilies are distinguished through homotypic interactions, where only domains within the same subfamily interact with each other. While death domain superfamily proteins are primarily involved in tumor necrosis factor (TNF) receptor-mediated or mitochondria-mediated apoptosis by promoting the activation and assembly of various signaling complexes, several member proteins also participate in other biological pathways and control various cellular processes, including inflammation, cell proliferation and migration, as well as insulin sensitivity. The Human genome encodes 37 DDs, 7 DEDs, 33 CARDs, and 22 PYDs [2, 3], making the death structural domain superfamily one of the largest superfamilies of mammalian proteins [4].

Viruses are also reported to possess death homologous sequences, with which viruses use to interfere with host apoptotic and inflammatory responses [5-8]. Because of their critical physiological functions *in vivo* and favorable physical properties (solubility, stability, polymorph, etc.) *in vitro*, the death structural domains have been extensively explored using NMR spectroscopic techniques. However, current literature lacks a comprehensive review on the NMR studies of death domain superfamily, and the purpose of this review is to fill the gap in the literature.

1.1. Death Receptor-Mediated Signaling Pathways: DD and DED Complexes

Apoptosis, or programmed cell death, is a naturally occurring cellular process to remove cells that are no longer needed or that will be detrimental to an organism or tissue in a neat and orderly manner, producing virtually no inflammatory responses. There are broadly defined "extrinsic" and "intrinsic" apoptotic pathways, in which death signals activate a family of cysteine proteases, named caspases, which act in a proteolytic cascade to dismantle and remove the dying cells [9]. TNFR-mediated pathways are one type of the extrinsic pathways, in which the cell death is determined by extracellular conditions.

Death receptors, constituting a subfamily of TNF receptor superfamily, are characterized by a cysteine-rich extracellular domain and a homologous cytosolic domain, known as DD. The death receptors, located at cell surface, receive instructive apoptotic signals initiated by specific death ligands, and subsequently activate death caspases within seconds of ligand binding, causing apoptotic demise of the cell within hours. Adapter molecules, like Fas-associated death domain (FADD), TNFR1-associated death domain (TRADD), or Daxx, contain both DDs to interact with the death receptor DDs and DEDs to recruit procaspases and transmit the apoptotic signal to the death-machinery, activating caspases [10].

The best-characterized death receptors are Fas and TNFR1. Fas, upon binding to its ligand, Fas Ligand (FasL), a homotrimeric protein, oligomerizes its DDs to bind adapter protein FADD. The FADD protein then binds via its DED to the homologous DED prodomain on procaspase-8. The complex of Fas, FADD, and procaspase-8 is termed death-inducing signaling complex (DISC). Upon recruited into the DISC, procaspase-8 activates through self-cleavage.

Active caspase-8 then activates downstream effector caspases-3, and -7, committing the cell to apoptosis [11]. TNFR1 activation initiates the formation of distinct complexes I and II in a temporal manner, and the balances between these complexes determines the survival or death of the cell [12]. A TNFR1 complex possessing the DD-containing protein TRADD, TNF receptor associated factor-2 (TRAF2), cellular inhibitor of apoptosis-1 (CIAP1), and the kinase receptor-interacting protein-1 (RIP-1) assembles at the plasma membrane within minutes after activation, termed Complex I, which subsequently recruits I-κB-kinase (IKK) and activates nuclear factor-κB (NF-κB), which induces the expression of IAP, an inhibitor of caspases-3, -7 and -9, leading to cell survival [13]. Complex II, formed after dissociation of the first one and insufficient NF-κB activation, recruits FADD and activates procaspases-8 or -10.

The effects between the two complexes are controlled by FLICE-like inhibitory proteins (FLIPs), a family of cellular and viral tandem DED-containing proteins that interact with FADD and block activation of caspases-8 and -10 in complex II (and DISC) [14].

1.2. Apoptosome and PIDDosome: CARD Complexes

The apoptosome is formed in response to various intracellular stress, including growth factor withdrawal, DNA damage, and unfolding stresses in the endoplasmic reticulum (ER), and is induced in a mitochondria-dependent manner, generally referred to as the intrinsic apoptosis pathway. Apaf-1 (apoptotic protease-activating factor 1), consisting of an N-terminal CARD, a central nucleotide-binding oligomerization domain (NOD), and a C-terminal tryptophan-aspartic acid (WD) repeat domain, mediates the formation of apoptosome for caspase activation [15, 16].

Cytochrome *c* normally resides at the intermembrane space of the mitochondria. Under stress, cytochrome *c* can be released from the mitochondria into the cytosol, where it binds to the Apaf-1 C-terminal WD domain, leading to an ATP- or dATP-dependent oligomerization of Apf-1 to form the apoptosome, which in turn recruits and activates caspase-9 via CARD-CARD interaction [17].

Under genotoxic stress, the DD-containing p53-induced protein with a death domain (PIDD) mediates the activation of procaspase-2, an evolutionarily conserved initiator caspase with a CARD prodomain [18], using the adapter molecule RIP-associated ICH-1/CED-3-homologous protein with a death domain (RAIDD), which possesses both DD and CARD domains [19], to recruit procapspase-2, forming a ternary complex, referred to as PIDDosome [20, 21]. PIDD, similar to TNFR1, can also recruit the DD-containing kinase RIP, which in turn recruits IKK to activate NF-κB, acting as a switch between life and death in response to DNA damage [22, 23].

1.3. Inflammasome: PYD Complexes

PYD represents the newest addition to the death structural domain superfamily [24]. Among the 22 human PYD-containing proteins, 14 of them are NALP (NACHT, leucine-rich repeat and PYD-containing) proteins, which are members of the NOD-like receptor (NLR) family responsible for detection of molecular signals from bacteria, microbial toxins, and pathogens [25, 26]. NALPs contain an N-terminal PYD, a central NOD and a leucine-rich repeat (LRR) domain, and a C-terminal CARD. The N-terminal PYD of NALP recruits the adaptor molecule, apoptosis-associated speck-like protein containing a CARD (ASC), which possesses a N-terminal PYD and a C-terminal CARD. The subsequent CARD-CARD interaction brings in caspases-1 and -5 to form the ~700 kDa inflammasome complex for caspase-1 activation [27-29]. Activated caspase-1 cleaves and activates cytokines of the IL-1 family, such as IL-1β, leading to NF-κB activation and elicitation of innate immunity [29].

2. NMR Structural and Dynamic Insights of Death and Inflammation Complexes

Based on available structures in the Protein Data Bank (PDB), the death domain superfamily was first established by NMR spectroscopy, from structures of Fas DD (PDB ID 1DDF) [30], FADD DED (PDB IDs 1A1Z and 1A1W) [31], RAIDD CARD (PDB ID 3CRD) [32], and NALP1 PYD (PDB ID 1PN5) [33]. X-ray crystallography soon joined in solving dozens of death domain superfamily structures.

Currently, there are 25, 11, 21, and 10 PDB entries for DD, DED, CARD, and PYD subfamilies, respectively, with an approximately equal number of crystal and solution structures (Table 1). However, almost all complex structures involving death domain superfamily reported so far have been solved by X-ray crystallography, imposing the limitation of NMR spectroscopy in studying protein complexes.

Nevertheless, NMR spectroscopy provides invaluable structural insights into the mechanisms of these protein domains in mediating various signaling pathways, and solution conditions are closer to the cellular environment where these proteins reside, making the solution structures more biologically relevant.

Table 1. Structures of death domain superfamily proteins

X-ray crystallography	NMR spectroscopy
Death domain (DD) subfamily	
1D2Z: Pelle/Tube DD complex 1WMG: UNC5H2 DD 2A9I: IRAK4 DD 2G7R: lymphoma translocation protein 1 DD 2O71: RAIDD DD 2OF5: PIDDosome core DD complex 2YVI: Ankyrin-1 DD 3EZQ: Fas/FADD DD Complex 3MOP: The ternary Death Domain complex of MyD88, IRAK4, and IRAK2 3OQ9: Fas/FADD DD assembly 4F42: Neurotrophin p75NTR intracellular domain (monomer) 4F44: Neurotrophin p75NTR intracellular domain (dimer)	1NGR: p75 low affinity neurotrophin receptor DD, residues 334-418 1DDF: Fas DD (average structure) 1FAD: FADD DD domain, residues 89-183 1E3Y: human FADD/MORT1 DD (average) 1E41: human FADD/MORT1 DD 1ICH: TNFR1 DD 1WH4: mouse IRAK4 DD 1YGO: Pelle DD 1WXP: nuclear matrix protein p84 DD 2GF5: Intact FADD (MORT1) 2D96: NF-κB DD 2DBF: NF-κB p105 subunit DD 2IB1: p45 DD 2YQF: Ankyrin-1 DD
Death effector domain (DED) subfamily	
2BBR: vFLIP MC159 2BBZ: vFLIP MC159 2F1S: vFLIP MC159 4IZ5: ERK2 phosphomimetic mutant/PEA-15 complex 4IZ7: Non phosphorylated ERK2/PEA-15 DED complex 4IZA: Dually phosphorylated ERK2/PEA-15 DED complex	1A1W: FADD DED, F25Y mutant (average) 1A1Z: FADD DED, F25G mutant (average) 1N3K: Full-length PEA-15 2GF5: Intact FADD (MORT1) 2LS7: High definition structure of PEA-15 DED
Caspase recruitment domain (CARD) subfamily	
1CY5: Apaf-1 CARD 2YGS: Apf1-1 CARD 3YGS: Apf1-1 CARD in complex with prodomain of procaspase-9	3CRD: RAIDD CARD domain 1C15: Apaf-1 CARD 1CWW: Apaf-1 CARD 1DGN: ICEBERG CARD

X-ray crystallography	NMR spectroscopy
Caspase recruitment domain (CARD) subfamily	
1Z6T: Apaf-1 bound to ADP 1A5Y: CED-4/CED-9 complex 2NSN: Nod1 CARD 2P1H: Apaf-1 CARD 2NZ7: Nod1 CARD 3KAT: Human NLRP1 CARD 3LQQ: CED-4 apoptosome 3LQR: CED-4/CED3 complex 3T6P: apo cIAP monomer 4E9M: Nod1 CARD domain swapped dimer 4I16: CARMA1 CARD 4IFP: human NLRP1 CARD 4IKM: CARD8 CARD 4IRL: Zebrafish GBP-NLRP1 like protein CARD	2B1W: Nod1 CARD 2DBD: Nod1 CARD 2NK6: Full lenth ASC 2L9M: cIAP CARD
Pyrin domain (PYD) subfamily	
3QF2: NALP3 PYD 4EWI: NLRP4 PYD	1PN5: NALP1 PYD 1UCP: human ASC PYD 2DBG: human Myeloid cell nuclear differentiation antigen (MNDA) PYD 2HM2: ASC2 2DO9: NALP10 PYD 2KM6: NLRP7 PYD 2L6A: NLRP12 PYD 2M5V: human NLRP10/PYNOD PYD

In this section, we will focus on solution NMR studies, in addition to complementary X-ray data, to reveal atomic details of the formation of various complexes in apoptosis and inflammation.

2.1. The DISC (DD and DED) and the Activation of Caspase-8/-10

Upon ligand activation, the DISC is formed between the death receptor Fas, the adaptor molecule FADD, and the procaspase-8 or 10 via DD-DD (Fas-FADD) and DED-DED (FADD-procaspase) interactions [34]. The solution NMR structure of Fas DD (PDB ID 1DDF), the founding member of the DD subfamily, revealed a novel fold consisting of six antiparallel, amphipathic α-helices [30]. Soon after, the FADD DD was also determined by NMR, and site-directed mutagenesis study suggested that charged residues in helices α2 and α3 are involved in interaction with Fas in an antiparallel fashion, in which FADD DD α2 interacts with Fas DD α3, and vice versa [35]. A subsequent NMR dynamics study on FADD DD indicated a mostly globular domain with uniform values of relaxation parameters and generalized order parameter, S^2, while an increased amide proton solvent exchange rate was observed for α3 helix, correlating with higher degree of solvent exposure of this helix [36].

These earlier NMR structural and dynamic studies on individual DDs did not provide much information on the DD-DD complex formation. The first DD-DD complex was crystalized between monomeric Pelle DD and Tube DD (PDB ID 1D2Z) [37]. This X-ray structure delineated an asymmetric dimer in which Pelle DD utilizes α4 helix and α4-α5 loop to interact with α1-α2 corner, α5-α6 loop, and α6 helix of Tube DD to form a linear, open-ended complex. Many charged residues are involved in the interactions. The binding interfaces within DD-DD complexes are further expanded by recently crystal structures of PIDDosome DD core complex consisting of seven RAIDD DDs and five PIDD DDs (PDB ID 2OF5) [38], Fas/FAD DD complex with four FADD DD bound to four Fas DD at acidic pH (PDB ID 3EZQ) [39], and asymmetric Fas/FAD DD complex composed of 5-7 Fas DD and 5 FADD DD at pH 8.0 (PDB ID 3OQ9) [40].

Arguably lack of physiological relevance, these higher oligomeric DD complexes, nevertheless, provided tremendous details on the potential binding interfaces between two DDs. The oligomeric PIDDosome core complex revealed three major types of DD interactions: type I interface is located between α1 and α4 of the first DD and α2 and α3 of the second DD; type II interface is formed between α4 helix and α4-α5 loop of the first DD and α5-α6 loop and α6 helix of the second DD; and type III interface involves α3 helix of the first DD and α1-α2 and α3-α4 loops of the second DD [38]. These interfaces cover the majority of DD surface, and mutations on all interfaces lead to disruption of complex assembly and caspase-2 activation. These DD-DD interfaces were also observed in oligomeric Fas/FADD DD complex, and mutations at type I and type III interfaces disrupt complex formation *in vitro* and *in vivo* [40].

The DED structures that are relevant to DISC formation are only available for FADD, while structures for the prodomains of procaspase-8 and -10 are still unknown. FADD DED structures of F25Y and F25G mutants were determined by NMR and are homologous to Fas DD [31]. These structures also revealed a hydrophobic region formed mostly by residues on α2 helix. This hydrophobic patch has later been shown to be conserved among most tandem DEDs, including the viral FLIP MC159 [41, 42] and PEA-15 (phosphoprotein enriched in astrocytes, 15 kDa) [43], and is essential in facilitating DED-DED interaction to recruit procaspase-8 or -10 to the DISC. A second prominent surface feature on the DEDs, termed charge triad, is a conserved hydrogen-bond network formed among three charged residues, an aspartic acid (D) or glutamic acid (E) in helix α2, and an arginine residue (R) and an aspartic acid (D) residue, separated by one other amino acid (x), located on α5-α6 loop and α6 helix. There is also a conserved leucine residue (L) immediately after the last aspartic acid. This charge-triad motif is often denoted as E/D-RxDL. This motif was first observed in the crystal structure of MC159 [41], and later substantiated in a high-definition NMR structure PEA-15 (PDB ID 2LS7) [44].

The full-length FADD NMR structure revealed N-terminal DED and C-terminal DD are aligned with each other in an orthogonal, tail-to-tail fashion [45]. The individual domains in the full-length structure are similar to the isolated FADD DED [31] and DD [35, 36]. The first helix of the DD in the full-length structure, however, is shifted in position comparing to the isolated domain, due to the connection between the DED and DD in the full-length protein. NMR relaxation and residual dipolar coupling (RDC) [46] evidence indicated that the two domains in the protein tumble together as one entity. In addition, the spatial relationship between the two domains was unambiguously defined by inter-domain NOEs (nuclear Overhauser effects). The DD-DED interface was established through the interactions between α4-α5 loop of the DED and α4-α5 loop of the DD, as well as the short inter-domain linker (residues 87-92).

The potential interface between Fas DD and FADD DD is composed of a patch of positively charged residues and a predominantly nonpolar surface in the vicinity of the α5-α6 loop of the FADD DD, with additional small basic patch on the opposite side of the DD[35, 47, 48]. The interface to binding procaspase-8 on the DED was identified to involve residues on helices α1 and α4 [44]. FADD self-association is also required for death-receptor mediated apoptosis [49, 50], and the self-association seemed to be mediated by F25 and adjunct residues between helices α2 and α3 [44]. On the other hand, mutations within the charge-triad RxDL residues could render a disordered DED, resulting a loss in structural integrity, and therefore this region may not mediate self-association of FADD.

Despite recent advances in the delineation of the DISC formation, the understanding of caspases-8 and -10 activation is still plagued by the lack of the prodomain structures of the caspases and unavailability of DED-DED complexes besides the tandem DED-DED structures of MC159 [41, 42]. We attributed the lack of DED-DED complexes to the unusual dynamic nature of DEDs mediated by the charge triad [51], limiting the ability of forming complex crystals.

In addition, there is still a mystic stoichiometry within the DISC: a trimeric Fas DD is formed upon ligand binding, while procaspase-8 and -10 activation requires a dimeric FADD and caspases-8 or -10. Although a hexamer Fas DD/FADD DD is proposed to resolve the stoichiometric issue [40], none of the higher oligomeric complexes currently available were crystallized as hexameric. Therefore, the higher oligomeric states of these DD-DD complexes are likely formed as a result of crystal packing, and do not have direct biological relevance, although these structures clearly illustrated the interactions between two DD domains.

2.2. The PIDDosome (DD and CARD) and the Activation of Caspase-2

The formation of PIDDosome involves DD and CARD. The core complex of PIDDosome between PIDD DD and RAIDD DD is crystalized in an oligomeric form (five PIDD DD and seven RAIDD) (PDB ID 2OF5) [38]. The adaptor molecule RAIDD also possesses a CARD, which it uses to recruit procaspase-2.

CARD domains are available as either the prodomains of procaspases or part of the adaptor molecules for the assembly of apoptotic and inflammatory signaling complexes. The first CARD structure was solved by NMR spectroscopy for RAIDD (PDB ID 3CRD) [32]. The structure shows a homologous topology to other DDs and DEDs, and is tightly folded into a six-helix bundle.

There is a unique surface feature of the CARD in that the surface is polarized with a basic and an acidic patch on opposite sides. The surface polarity seems to be conserved across all CARD structures, and may be used for CARD-CARD interactions. There is currently no structural information on the prodomain of procaspase-2, although the interaction could be similar to

Apaf-1 CARD and procaspase-9 CARD in the formation of the apoptosome, which is detailed in the next section.

2.3. The Apoptosome (CARD) and the Activation of Caspase-9

The apoptosome is formed when cytochrome *c* is released from mitochondria. Upon binding to cytochrome *c* and ATP/dATP, Apaf-1 undergoes a major conformational change, exposing its CARD to recruit and activate procaspase-9. Apaf-1 CARD structure was solved by both X-ray crystallography (PDB IDs 1CY5 [52] and 2YGS [53]) and NMR spectroscopy (PDB IDs 1C15 [54] and 1C1W [55]), together with a complex structure between Apaf-1 CARD and the prodomain of caspase-9 (PDB ID 3YGS [53]). Apaf-1 CARD shares the same structural fold as other death domain superfamily members, but a kink in the middle of the N-terminal helix was observed in both solution NMR and crystal structures, which breaks the helix into α1a and α1b [53-55]. The caspase-9 binding interface was mapped to the acidic surface of Apaf-1 CARD around helices α2 and α3 [54, 55]. The surface charge polarity was also confirmed in Apaf-1 CARD [53]. The procaspase-9 CARD in the complex with Apaf-1 CARD adopts a similar global fold, with first helix breaks into α1a and α1b [53]. The interface between the two CARDs is formed by complementary concaved surfaces, composed of the positively charged helices α1a/α1b and α4 on the procaspase-9 CARD and negatively charged helices α2 and α3 on the Apaf-1 CARD [53]. The formation of the complex seems to be driven by both electrostatic and hydrogen bonding interactions and hydrophobic effects. The crystal structure of Apaf-1 bound to ADP (PDB ID 1Z6T) revealed the mechanism how hydrolysis of ATP/dATP exposes the binding surface to procaspase-9 [56].

2.4. Inflammation Responses (CARD and PYD) and the Activation of Caspase-1

The CARD-containing proteins, Nods, are responsible for detection of pathogen-associated molecular patterns (PAMPs) and eliciting innate immunity against these pathogens. Nod1 consists of an N-terminal CARD, a central NOD, and a C-terminal LRR domain [57]. Nod1 uses the LRR to recognize ligands on intracellular bacteria, oligomerizes through the NOD, and recruits the kinase RIP2 (RICK) via CARD-CARD interactions [58]. RIP2 can either recruit the IKK complex for NF-κB activation [59], or activate caspase-1 through the CARD-CARD interaction to subsequently cleave and activate pro-IL-1β [60]. The Nod1 CARD structure was determined by NMR spectroscopy (PDB ID 2B1W), and shows similar folds as other CARDs, while the α6 helix displays an unusual length and position [61]. The Nod1-RIP2 interaction seems to follow the same electrostatic pattern as other CARD-CARD interactions in that the acidic residues on Nod1 interact with basic residues on RIP2. A dimeric form of Nod1 CARD was also crystalized with α6 helix swapped between the two monomers (PDB IDs 2NSN [62] and 2NZ7 [63]).

While this domain swapped dimer of Nod1 CARD may suggest another level of regulation and activation of NLR proteins, the physiological relevance is unclear. The monomer/dimer transition is pH dependent that the Nod1 CARD dimer is formed at pH 6.0 and below, and stabilized by an interchain disulfide bond, while it stays as monomer at pH 7.0 and above [63].

The Nod1 pathway for the activation of IL-1β can be inhibited by caspase-1 inhibitor ICEBERG, a protein with a single CARD, which interacts with procaspase-1 and prevents its association with RIP2 [64]. The solution NMR structure of ICEBERG displays an antiparallel six-helix bundle with Greek-key topology, similar to other CARDs. Helix α1 is interrupted by a kink (α1a and α1b), and helix α6 is a 3_{10} helix (PDB ID 1DGN) [64]. ICEBERG also possesses separate charged patches, with positively charged patch formed by basic residues on α1, α4, and α6 on one surface, and negatively charged patch formed by acidic residues on α2 and α5 on the opposite surface. A third charged patch is comprising acidic residues from α4-α5 loop and α5 helix. The ICEBERG/procaspase-1 interaction is also suggested to be mediated by charge-charge interactions, similar to other CARD-CARD interactions.

Another pathway for procaspase-1 activation is the NLRP-mediated inflammasome complex. Human NLRP1 possesses both an N-terminal PYD and a C-terminal CARD. The NLRP1 CARD crystal structure (PDB ID 4IFP) is similarly folded as other known CARDs, and exhibits charged surface patches [65].

The N-terminal PYD structure of NLRP1 was also available and was determined by NMR spectroscopy (PDB ID 1PN5) [33]. The distinction between PYD and other death domain superfamily structures is that the α3 helix is replaced by a flexibly disordered loop. There is also a large negatively charged patch on the surface of the PYD, which is conserved among PYD subfamily members. Various other PYD structures of NLRP proteins have been solved by both NMR and crystallography. NLRP7 PYD (PDB ID 2KM6) adopts a six-helix bundle with a stabilized α3 helix and a well ordered α2-α3 loop [66], in contrast to a flexible loop as shown in NLRP1 PYD [33]. NLRP3 crystal structure (PDB ID 3QF2) is a symmetric dimer with each monomer forming a canonical six-helix bundle [67], while the dimer may be formed coincidentally by crystal packing. The non-inflammasome NLRP12 PYD adopts a typical six-helix bundle death domain fold (PDB ID 2L6A) as revealed by NMR spectroscopy, and this PYD involves in a non-homotypic interaction with Fas-associated factor 1 (FAF-1) [68]. All residues participating in the NLRP12 PYD/FAF-1 interaction come from helices α2-α3, distinct from other NLRP PYD structures. NLRP4 PYD folds similarly to other PYDs, but helix α3 and α2-α3 loop are unique in this crystal structure (PDB ID 4EWI) [69].

It also does not seem to bind to the adaptor protein ASC [69]. NLRP10, a unique NLRP member that lacks LRR, is shown to inhibit ASC-mediated procaspase-1 processing. Solution NMR structure of NLRP10 PYD (PDB ID 2M5V) displays a canonical six-helix bundle, with a stabilized helix α3 [70], although another NLRP10 PYD structure shows a disordered loop in place of helix α3 (PDB ID 2DO9).

The adaptor molecule between NLRPs and procaspase-1, ASC, possesses an N-terminal PYD to interact with NLRP PYDs and a C-terminal CARD to recruit procaspase-1. The ASC PYD structure determined using solution NMR spectroscopy (PDB ID 1UCP) shows that although the PYD also folds into the six-helix bundle as other death domain superfamily members, the inter-helical angles, lengths of the helices and inter-helical loops are variable [71]. The PYD surface also exhibits a charge polarity, with most positively charged residues located in helices α2 and α3, while most negatively charged residues located in helices α1 and α4. The distinct charge patches on PYD surface suggests an important role of electrostatic interaction in PYD-PYD association [71], analogous to CARD-CARD interactions [53]. A related PYD-only protein, ASC2, shows a characteristic six-helix bundle fold, with an exceptionally long α2-α3 loop (PDB ID 2HM2) [72], which was also seen in the ASC PYD structure. The full-length ASC structure exhibits structurally independent PYD and CARD connected by a 23-residue linker (PDB ID 2KN6) [73].

The two domains do not seem to interact with each other, as no inter-domain NOEs and NOEs with linker were observed. The full-length ASC PYD shares general characteristics with other know PYD structures, including the long loop between helices α2 and α3, and is structurally similar to isolated ACS PYD and ASC2, suggesting that the presence of the CARD does not perturb the PYD structure. The ASC CARD, however, shows some structural peculiarities compared with other known CARDs. The ASC CARD does not seem to have helix α1a, and helices α2 and α3 also adopt a different orientation, as compared to Apaf-1 CARD [73]. In addition, ASC CARD does not have the two extensive oppositely charged patches as shown on Apaf-1 CARD, in contrast, the positive and negative charges are evenly spread out throughout the ASC CARD surface [73].

Protein dynamics plays substantial roles in protein functions and interactions. In addition to their structures, NMR dynamic information was obtained for various NLRP PYDs and ASC/ASC2. The most prominent dynamic discrepancy among PYDs is within the unusually long α2-α3 loop and the α3 helix, as reported by the auto-correlated ^{15}N-relaxation parameter, the heteronuclear ^{15}N [74] NOE, which illustrates backbone motions on fast *ps – ns* timescale. NLRP1 PYD exhibits particularly low heteronuclear ^{15}N [74] NOE between helices α2 and α4, indicating a reduced effective rotational correlation time for residues in this region compared to other parts of the protein [33]. However, no significantly reduced heteronuclear $^{15}N\{^1H\}$ NOE were observed for other PYDs that possess rigid α2-α3 loop and well-ordered α3 helix, including ASC PYD [71], ASC2 PYD [72], full-length ASC [73], NLRP7 PYD [66], and NLRP10 PYD [70]. In fact, in these proteins, the region between helices α2 and α4 exhibits rigidities nearly identical to their neighboring helices. Although NLRP12 PYD displays a small increase in backbone flexibility in α2-α3 loop compared to NLRP7 and NLRP10 PYD, it is much distinct from NLRP1 PYD in the backbone dynamics [68]. However, some residues located in the α2-α3 loop exhibit increased motions on a slower timescale than those of the rest of the protein [66, 72].

Among the four subfamilies, DD subfamily has five known DD complex structures (PDB IDs 1D2Z, 2OF5, 3EZQ, 3MOP, 3OQ9), and CARD subfamily has one CARD complex structure (PDB ID 3YGS). DED subfamily has one structure with tandem DEDs (PDB ID 2BBR). These complex structures provide valuable insight into the homotypic interactions of death domain superfamily. PYD subfamily, however, has currently no complex or tandem structures, limiting the understanding of the potential PYD-PYD interactions, although the interactions have been suggested to be electrostatic, homologous to CARD-CARD interactions. Various types of potential PYD-PYD interaction have been proposed [75]. Nevertheless, the limited information on PYD homotypic interaction and the discrepancy on the PYD fold on certain proteins, particularly in regions of the α2-α3 loop and α3 helix, hinder the understanding of formation of inflammasome and the activation of caspase-1, which is an essential step for innate immune responses.

3. Beyond Homotypic Interactions

Although most interactions in the death domain superfamily are homotypic, there are cases that members of death domain superfamily involve in heterotypic interactions, particularly in the DED subfamily (see a recent review [76]). In this chapter, we will focus on one DED-containing protein, PEA-15 [77], and summarize the most recent advances in the NMR studies of this multifunctional small protein in heterotypic protein-protein interactions.

PEA-15 is small cytosylic protein consisting of 130 amino acids (~15 kDa). It contains an N-terminal DED and a C-terminal irregularly structured tail.

Ubiquitously expressed in different human tissues and highly conserved among mammals [78, 79], PEA-15 interacts with a diverse array of proteins with and without DEDs [76], and its binding specificity is mediated by the phosphorylation on two serine residues on the C-terminal tail [80, 81], S^{104}, the substrate for protein kinase C (PKC), and S^{116}, phosphorylated by calcium/calmodulin-dependent protein kinase II (CamKII) or protein kinase B/Akt [82-84]. The N-terminal DED of PEA-15 can interact with other DED-containing proteins, including FADD, FADD-like IL-1β-converting enzyme (FLICE), and procaspase-8 [85], and it can be recruited to the DISC [86], inhibiting extrinsic apoptotic pathway [87]. PEA-15 can also restrains cellular proliferation by regulating mitogen-activated protein (MAP) kinase signaling pathway through binding to extracellular regulated kinase 1 and 2 (ERK1/2), preventing translocation of ERK1/2 into the nucleus, and blocking ERK dependent cell proliferation and migration [88, 89]. In addition, PEA-15 is overexpressed in adipose and skeletal muscle tissues and in skin fibroblasts from type 2 diabetic individuals independent of obesity, causing resistance to insulin action in glucose uptake [90]. PEA-15 interacts with phospholipase D (PLD) isoforms 1 and 2, elevating intracellular levels of diacylglycerol, thereby activating diacylglycerol-sensitive PKCα isoform [91]. Activation of PKCα in PEA-15 overexpressing cells and tissues prevents insulin induction of the PKCζ isoform, which is the major activator of glucose transporter 4 (GLUT4) [92, 93]. Because of PEA-15's roles in diabetes, it is also referred to as phosphoprotein enriched in diabetes, PED-15. Figure 1 summarizes these biological pathways that PEA-15 is involved.

3.1. PEA-15/ERK2 Interaction

PEA-15 structure was first determined by solution NMR spectroscopy (PDB ID 1N3K), and its interaction with ERK2 was suggested from chemical shift perturbation (CSP) and site-directed mutagenesis [43]. The NMR backbone dynamics was also reported for PEA-15 in its free form [94].

Based on this NMR model and mutagenesis study, it was proposed that residues from both DED and the C-terminal tail were essential in binding ERK2, and the charge-triad of PEA-15 DED, particularly the D^{74} residue, was involved in the binding interface.

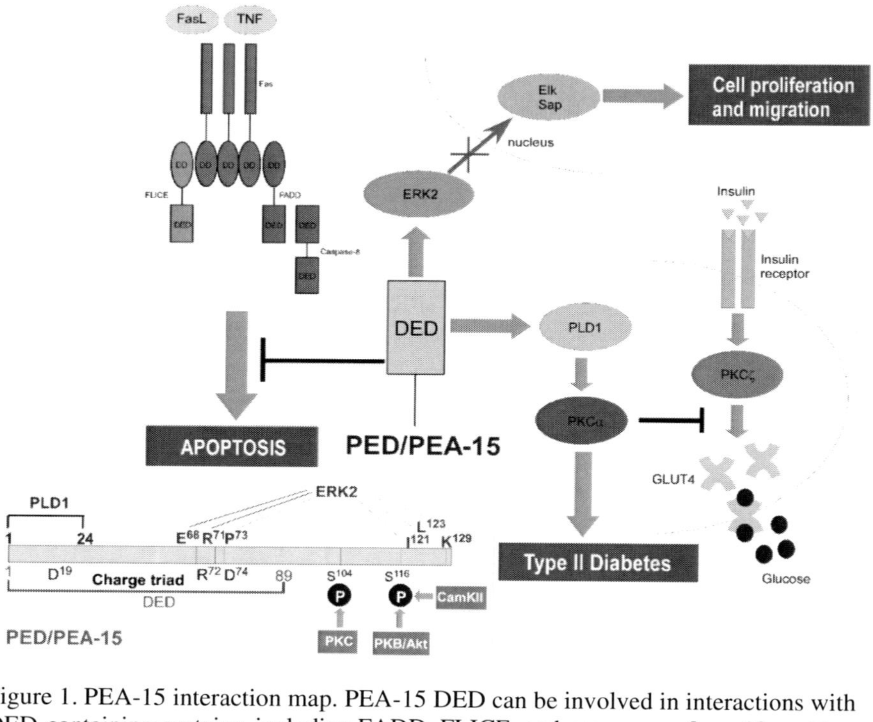

Figure 1. PEA-15 interaction map. PEA-15 DED can be involved in interactions with DED-containing proteins, including FADD, FLICE, and procaspase-8 or -10, and be recruited to the DISC, inhibiting apoptosis. PEA-15 can also translocate to ERK1/2 from nucleus into cytosol, preventing ERK1/2-depedent cell proliferation and migration. PEA-15 overexpression and interaction with PLD1 may lead to type II diabetes. The lower left insert shows PEA-15 sequence features, including residues 1-89 for DED, charge-triad residues D^{19}-R^{72}-D^{74}L (black lines and labels), phosphorylation sites (green lines and labels) and corresponding kinases, ERK2 binding residues (purple lines and labels), and PLD1 interaction segment (residues 1-24).

The involvement of both DED and C-terminal tail of PEA-15 in binding ERK2 was later confirmed by fluorescence binding assay [89, 95], and a dynamic light scattering study showed that PEA-15 and ERK2 forms a strong 1:1 complex [96].

We additionally reported CSP using TROSY (transvers relaxation optimized spectroscopy) technique and measured NMR relaxation parameters, including the longitudinal relaxation rate R_1, transverse relaxation rate R_2, and auto-correlation relaxation $^{15}N\{^1H\}$ NOE, for the full-length PEA-15 in ERK2-bound form, at 800 MHz field [51]. The TROSY CSP and the backbone dynamics also confirmed the involvement of C-terminal residues in binding ERK2, while the direct involvement of charge-triad residue, D^{74}, at the binding interface was questionable. In fact, all charge-triad residues, D^{19}-R^{72}-D^{74}L, displayed minimum CSPs, and no special dynamic property changes were observed for these residues in the free and ERK2-bound forms when compared with majority of other DED residues. On the other hand, the most prominent CSPs occurred among residues in helices α5 and α6. In addition, reduced spectral density mapping of the relaxation parameters indicated that overly complicated backbone dynamics, particularly for the motions at a slow $ms - s$ timescales, exhibits in the PEA-15 DED in the complex. Probably the most surprising result from this dynamic study was the overall rotational correlation time of the PEA-15/ERK2 complex was determined to be 11.6 ns, much smaller than the expected value for a 57-kDa complex. We, therefore, did a calculation of the per residue correlation times, which astonishingly showed that PEA-15 DED tumbles at two distinct correlation times, a fast rotational correlation time of ~11.5 ns, exhibiting mostly in residues from helices α2, α3, and α4, and a slow correlation time of ~15 ns, exhibiting mostly in residues from helices α1, α5, and α6. The slow correlation time is actually closer to expected value of a 57-kDa complex [51]. Based on this study, we proposed a new model for PEA-15 and ERK2 interaction. In this model, the C-terminal tail recognizes the ERK2 and subsequently induces a substantial conformational change of DED so that the residues in helices α1, α5, and α6 engage in binding ERK2, while helices α2, α4, and α5 do not interact with ERK2, and tumble relatively freely from the rest of the complex. The charge-triad residues, D^{19}-R^{72}-D^{74}L, located at a hinge position between these two dynamically distinct segments, mediate the necessary conformational changes to accommodate ERK2 binding (Figure 2).

Although we proposed a new role for the charge-triad in mediating ERK2 binding, the original NMR structure (1N3K) did not delineate the hydrogen-bonding network, which was clearly visible in the MC159 crystal structure

(2BBR). The lack of electrostatic and hydrogen bonding interactions in most NMR structures is mostly due to the simplification of NMR structural calculations and limited NOE restraints among polar and charged residues. We developed a new structural refinement scheme that incorporates residual dipolar couplings (RDCs) measured from multiple media and explicit solvent protocol.

Using this new refinement scheme, we determined the PEA-15 DED structure (PDB ID 2LS7), which allowed us to identify various electrostatic and hydrogen bonding interactions, including the charge-triad, on the protein surface [44]. We have termed our structure that is refined using dual-medium RDC data and explicit solvent protocol and has quality and details close to crystal structures as "high-definition" NMR structure.

We also tried to assess the substantial conformational change in our proposed PEA-15/ERK2 binding model using RDC data for the free and ERK2-bound forms. The RDC data confirmed our observation from NMR dynamics that PEA-15 DED behaves as two distinct segments upon binding to ERK2.

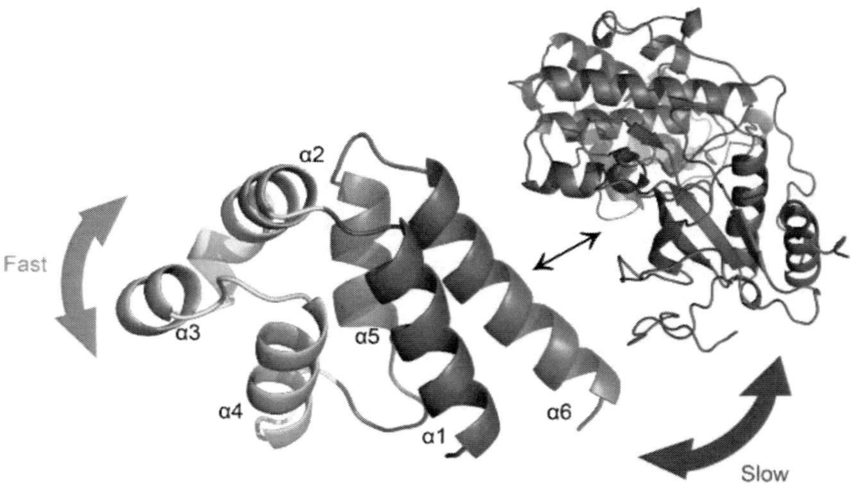

Figure 2. PEA-15 interaction model with ERK2. PEA-15 binds to ERK2 using residues from helices α1, α5, and α6, and this segment tumbles together with ERK2 at a slow rotational correlation time of ~15 ns. The other segment, composing of helices α2, α3, and α4, undergoes a major conformational change and tumble relatively freely from other parts of the complex, and possesses a fast rotational correlation time of ~11.5 ns. The conformational change seems to be mediated by the charge-triad hydrogen bonding network among residues D^{19}-R^{72}-D^{74}, located at a hinge position between the fast- and slow-tumbling segments of the DED.

In addition, the RDC data clearly illustrated that slow-tumbling segment consisting of helices α1, α5, and α6 undergoes little conformational change, although this segment directly interacts with ERK2, while the fast-tumbling segment composed of helices α2, α3, and α4 exhibits most conformational reorientations and complicated protein backbone flexibility at both fast $ps - ns$ and slow $ms - s$ timescales [97]. Recent crystal structures of PEA-15/ERK2 complexes confirmed our model in that charge-triad residues do not directly involve in binding ERK2, while a few other residues from helices α5 and α6 are engaged in the binding interface [98]. In addition, the crystal structures have missing residues from the α2-α3 loop and the α3 helix due to the lack of electron density, consistent with our NMR dynamics that this region displays complex backbone motions and flexibility. This observation in PEA-15 DED is homologous to some PYD proteins in that the α2-α3 loop and α3 helix could be disordered in either free or complex forms, and may engage in more roles in assisting protein complex formationn.

3.2. PEA-15/PLD1 Interaction

PLD1 is another major non-homotypic interaction partner for PEA-15. PLD1 interacts with PEA-15 using its C-terminal domain, termed D4 [91], and most of the interaction occurs at the N-terminal region of D4, termed D4α [99]. PEA-15 binding interface seemed to reside in N-terminal residues 1-24 [100].

A recent NMR study of PEA-15/PLD1-D4α suggested residues from both PEA-15 DED and C-terminal tail are involved in PLD1 interaction. The DED residues defining the binding surface are mostly located in α1, α3, α4 helices and α1-α2 and α3-α4 loop, and residues in α5 helix may be involved in allosteric regulation of the binding [101]. This study, together with our PEA-15/ERK2 binding studies, supported the mutual exclusive binding modes between PLD1 and ERK2 [99], as the PEA-15 binding interfaces for ERK2 and PLD1 are partially superimposable.

As seen in the PEA-15 complexes with ERK2 and PLD1, it seems to be essential for DEDs to maintain proper flexibility to adopt various conformations in order to accommodate distinct proteins of different shapes and sizes and to promote protein-protein interactions. Key electrostatic and hydrogen bonding interactions on the DED surface may be central to facilitate and mediate protein-protein interactions by maintaining the conformational flexibility and positioning key residues into optimal orientations to interact with other proteins. Disruption of these key interaction networks could reduce the flexibility of DED, resulting in a loss of ability to interact with other proteins.

The conformational changes and protein dynamics need to be taken into consideration if such protein-protein interactions involving DEDs are targeted therapeutically.

REFERENCES

[1] Park, H. H., et al., The Death Domain Superfamily in Intracellular Signaling of Apoptosis and Inflammation. *Annual Review of Immunology,* 2007. 25(1): p. 561-586.

[2] Kwon, D., et al., A comprehensive manually curated protein-protein interaction database for the Death Domain superfamily. *Nucleic Acids Res.*, 2012. 40(Database issue): p. D331-6.

[3] Reed, J. C., K. S. Doctor, and A. Godzik, The domains of apoptosis: a genomics perspective. *Sci. STKE*, 2004. 2004(239): p. re9.

[4] McEntyre, J. R. and T. J. Gibson, Patterns and clusters within the PSM column in TiBS, 1992-2004. *Trends Biochem. Sci.,* 2004. 29(12): p. 627-33.

[5] Hu, S., et al., A novel family of viral death effector domain-containing molecules that inhibit both CD-95- and tumor necrosis factor receptor-1-induced apoptosis. *J. Biol. Chem.,* 1997. 272(15): p. 9621-4.

[6] Bertin, J., et al., Death effector domain-containing herpesvirus and poxvirus proteins inhibit both Fas- and TNFR1-induced apoptosis. *Proc. Natl. Acad. Sci. U S A,* 1997. 94(4): p. 1172-6.

[7] Thome, M., et al., Viral FLICE-inhibitory proteins (FLIPs) prevent apoptosis induced by death receptors. *Nature,* 1997. 386(6624): p. 517-21.

[8] Johnston, J. B., et al., A poxvirus-encoded pyrin domain protein interacts with ASC-1 to inhibit host inflammatory and apoptotic responses to infection. *Immunity*, 2005. 23(6): p. 587-98.

[9] Czerski, L. and G. Nunez, Apoptosome formation and caspase activation: is it different in the heart? *J. Mol. Cell Cardiol.*, 2004. 37(3): p. 643-52.

[10] Zimmermann, K. C., C. Bonzon, and D. R. Green, The machinery of programmed cell death. *Pharmacol. Ther.*, 2001. 92(1): p. 57-70.

[11] Pirnia, F., et al., Mitomycin C induces apoptosis and caspase-8 and -9 processing through a caspase-3 and Fas-independent pathway. *Cell Death Differ.*, 2002. 9(9): p. 905-14.

[12] Sutheesophon, K., et al., Involvement of the tumor necrosis factor (TNF)/TNF receptor system in leukemic cell apoptosis induced by histone deacetylase inhibitor depsipeptide (FK228). *J. Cell Physiol.*, 2005. 203(2): p. 387-97.

[13] Swanton, E., et al., Bcl-2 regulates a caspase-3/caspase-2 apoptotic cascade in cytosolic extracts. *Oncogene*, 1999. 18(10): p. 1781-7.

[14] Thome, M. and J. Tschopp, Regulation of lymphocyte proliferation and death by FLIP. *Nat. Rev. Immunol.*, 2001. 1(1): p. 50-8.

[15] Zou, H., et al., Apaf-1, a human protein homologous to C. elegans CED-4, participates in cytochrome c-dependent activation of caspase-3. *Cell*, 1997. 90(3): p. 405-13.

[16] Li, P., et al., Cytochrome c and dATP-dependent formation of Apaf-1/caspase-9 complex initiates an apoptotic protease cascade. *Cell*, 1997. 91(4): p. 479-89.

[17] Adams, J. M. and S. Cory, Apoptosomes: engines for caspase activation. *Curr. Opin. Cell Biol.*, 2002. 14(6): p. 715-20.

[18] Lassus, P., X. Opitz-Araya, and Y. Lazebnik, Requirement for caspase-2 in stress-induced apoptosis before mitochondrial permeabilization. *Science*, 2002. 297(5585): p. 1352-4.

[19] Duan, H. and V. M. Dixit, RAIDD is a new 'death' adaptor molecule. *Nature*, 1997. 385(6611): p. 86-9.

[20] Tinel, A. and J. Tschopp, The PIDDosome, a Protein Complex Implicated in Activation of Caspase-2 in Response to Genotoxic Stress. *Science*, 2004. 304(5672): p. 843-846.

[21] Berube, C., et al., Apoptosis caused by p53-induced protein with death domain (PIDD) depends on the death adapter protein RAIDD. *Proc. Natl. Acad. Sci. U S A*, 2005. 102(40): p. 14314-20.

[22] Janssens, S., et al., PIDD mediates NF-kappaB activation in response to DNA damage. *Cell*, 2005. 123(6): p. 1079-92.

[23] Cuenin, S., et al., p53-induced protein with a death domain (PIDD) isoforms differentially activate nuclear factor-kappaB and caspase-2 in response to genotoxic stress. *Oncogene*, 2008. 27(3): p. 387-96.

[24] Kohl, A. and M. G. Grutter, Fire and death: the pyrin domain joins the death-domain superfamily. *C. R. Biol.*, 2004. 327(12): p. 1077-86.

[25] Tschopp, J., F. Martinon, and K. Burns, NALPs: a novel protein family involved in inflammation. *Nat. Rev. Mol. Cell Biol.*, 2003. 4(2): p. 95-104.

[26] Benko, S., D. J. Philpott, and S. E. Girardin, The microbial and danger signals that activate Nod-like receptors. *Cytokine,* 2008. 43(3): p. 368-73.

[27] Martinon, F., K. Burns, and J. Tschopp, The inflammasome: a molecular platform triggering activation of inflammatory caspases and processing of proIL-beta. *Mol. Cell*, 2002. 10(2): p. 417-26.

[28] Franchi, L., et al., The inflammasome: a caspase-1-activation platform that regulates immune responses and disease pathogenesis. *Nat. Immunol.*, 2009. 10(3): p. 241-7.

[29] Schroder, K. and J. Tschopp, The inflammasomes. *Cell,* 2010. 140(6): p. 821-32.

[30] Huang, B., et al., NMR structure and mutagenesis of the Fas (APO-1/CD95) death domain. *Nature,* 1996. 384(6610): p. 638-41.

[31] Eberstadt, M., et al., NMR structure and mutagenesis of the FADD (Mort1) death-effector domain. *Nature,* 1998. 392(6679): p. 941-5.

[32] Chou, J. J., et al., Solution structure of the RAIDD CARD and model for CARD/CARD interaction in caspase-2 and caspase-9 recruitment. *Cell,* 1998. 94(2): p. 171-80.

[33] Hiller, S., et al., NMR structure of the apoptosis- and inflammation-related NALP1 pyrin domain. *Structure,* 2003. 11(10): p. 1199-205.

[34] Wajant, H., The Fas signaling pathway: more than a paradigm. *Science,* 2002. 296(5573): p. 1635-6.

[35] Jeong, E. J., et al., The solution structure of FADD death domain. Structural basis of death domain interactions of Fas and FADD. *J. Biol. Chem.,* 1999. 274(23): p. 16337-42.

[36] Berglund, H., et al., The three-dimensional solution structure and dynamic properties of the human FADD death domain. *J. Mol. Biol.*, 2000. 302(1): p. 171-88.

[37] Xiao, T., et al., Three-dimensional structure of a complex between the death domains of Pelle and Tube. *Cell,* 1999. 99(5): p. 545-55.

[38] Park, H. H., et al., Death Domain Assembly Mechanism Revealed by Crystal Structure of the Oligomeric PIDDosome Core Complex. *Cell,* 2007. 128(3): p. 533-546.

[39] Scott, F. L., et al., The Fas-FADD death domain complex structure unravels signalling by receptor clustering. *Nature,* 2009. 457(7232): p. 1019-1022.

[40] Wang, L., et al., The Fas-FADD death domain complex structure reveals the basis of DISC assembly and disease mutations. *Nat. Struct. Mol. Biol.,* 2010. 17(11): p. 1324-1329.

[41] Yang, J. K., et al., Crystal Structure of MC159 Reveals Molecular Mechanism of DISC Assembly and FLIP Inhibition. *Molecular Cell,* 2005. 20(6): p. 939-949.

[42] Li, F.-Y., et al., Crystal Structure of a Viral FLIP. *Journal of Biological Chemistry*, 2006. 281(5): p. 2960-2968.

[43] Hill, J. M., et al., Recognition of ERK MAP kinase by PEA-15 reveals a common docking site within the death domain and death effector domain. *EMBO J.*, 2002. 21(23): p. 6494-6504.

[44] Twomey, E. C. and Y. Wei, High-definition NMR structure of PED/PEA-15 death effector domain reveals details of key polar side chain interactions. *Biochemical and Biophysical Research Communications,* 2012. 424(1): p. 141-146.

[45] Carrington, P. E., et al., The Structure of FADD and Its Mode of Interaction with Procaspase-8. *Molecular Cell,* 2006. 22(5): p. 599-610.

[46] Braddock, D. T., et al., Rapid Identification of Medium- to Large-Scale Interdomain Motion in Modular Proteins Using Dipolar Couplings. *Journal of the American Chemical Society,* 2001. 123(35): p. 8634-8635.

[47] Bang, S., et al., Fas- and Tumor Necrosis Factor-mediated Apoptosis Uses the Same Binding Surface of FADD to Trigger Signal Transduction: A Typical Model For Convergent Signal Transduction. *Journal of Biological Chemistry,* 2000. 275(46): p. 36217-36222.

[48] Hill, J. M., et al., Identification of an expanded binding surface on the FADD death domain responsible for interaction with CD95/Fas. *Journal of Biological Chemistry,* 2004. 279(2): p. 1474-1481.

[49] Muppidi, J. R., et al., Homotypic FADD interactions through a conserved RXDLL motif are required for death receptor-induced apoptosis. *Cell Death Differ.*, 2006. 13(10): p. 1641-50.

[50] Sandu, C., et al., FADD self-association is required for stable interaction with an activated death receptor. *Cell Death Differ.,* 2006. 13(12): p. 2052-61.

[51] Twomey, E. C., D. F. Cordasco, and Y. Wei, Profound conformational changes of PED/PEA-15 in ERK2 complex revealed by NMR backbone dynamics. *Biochimica et Biophysica Acta (BBA) - Proteins and Proteomics,* 2012. 1824(12): p. 1382-1393.

[52] Vaughn, D. E., et al., Crystal structure of apaf-1 caspase recruitment domain: an α-helical greek key fold for apoptotic signaling. *Journal of Molecular Biology,* 1999. 293(3): p. 439-447.

[53] Qin, H., et al., Structural basis of procaspase-9 recruitment by the apoptotic protease-activating factor 1. *Nature,* 1999. 399(6736): p. 549-57.

[54] Zhou, P., et al., Solution structure of Apaf-1 CARD and its interaction with caspase-9 CARD: a structural basis for specific adaptor/caspase interaction. *Proc. Natl. Acad. Sci. U S A,* 1999. 96(20): p. 11265-70.

[55] Day, C. L., et al., Solution structure and mutagenesis of the caspase recruitment domain (CARD) from Apaf-1. *Cell Death Differ.,* 1999. 6(11): p. 1125-32.

[56] Riedl, S. J., et al., Structure of the apoptotic protease-activating factor 1 bound to ADP. *Nature*, 2005. 434(7035): p. 926-933.

[57] Inohara, N., et al., Nod1, an Apaf-1-like activator of caspase-9 and nuclear factor-kappaB. *J. Biol. Chem.,* 1999. 274(21): p. 14560-7.

[58] Inohara, N., et al., RICK, a novel protein kinase containing a caspase recruitment domain, interacts with CLARP and regulates CD95-mediated apoptosis. *J. Biol. Chem.,* 1998. 273(20): p. 12296-300.

[59] Inohara, N., et al., An induced proximity model for NF-kappa B activation in the Nod1/RICK and RIP signaling pathways. *J. Biol. Chem.*, 2000. 275(36): p. 27823-31.

[60] Thome, M., et al., Identification of CARDIAK, a RIP-like kinase that associates with caspase-1. *Curr. Biol.*, 1998. 8(15): p. 885-8.

[61] Manon, F., et al., Solution Structure of NOD1 CARD and Mutational Analysis of its Interaction with the CARD of Downstream Kinase RICK. *Journal of Molecular Biology*, 2007. 365(1): p. 160-174.

[62] Coussens, N. P., et al., Crystal structure of the Nod1 caspase activation and recruitment domain. *Biochemical and Biophysical Research Communications,* 2007. 353(1): p. 1-5.

[63] Srimathi, T., et al., Monomer/Dimer Transition of the Caspase-Recruitment Domain of Human Nod1†,‡. *Biochemistry,* 2008. 47(5): p. 1319-1325.

[64] Humke, E. W., et al., ICEBERG: a novel inhibitor of interleukin-1beta generation. *Cell,* 2000. 103(1): p. 99-111.

[65] Jin, T., et al., Structure of the NLRP1 caspase recruitment domain suggests potential mechanisms for its association with procaspase-1. *Proteins: Structure, Function, and Bioinformatics,* 2013. 81(7): p. 1266-1270.

[66] Le, H. T. and J. A. Harton, Pyrin- and CARD-only Proteins as Regulators of NLR Functions. *Front Immunol.*, 2013. 4: p. 275.
[67] Bae, J. Y. and H. H. Park, Crystal Structure of NALP3 Protein Pyrin Domain (PYD) and Its Implications in Inflammasome Assembly. *Journal of Biological Chemistry,* 2011. 286(45): p. 39528-39536.
[68] Pinheiro, A. S., et al., The NLRP12 Pyrin Domain: Structure, Dynamics, and Functional Insights. *Journal of Molecular Biology*, 2011. 413(4): p. 790-803.
[69] Eibl, C., et al., Structural and Functional Analysis of the NLRP4 Pyrin Domain. *Biochemistry,* 2012. 51(37): p. 7330-7341.
[70] Su, M.-Y., et al., Three-Dimensional Structure of Human NLRP10/PYNOD Pyrin Domain Reveals a Homotypic Interaction Site Distinct from Its Mouse Homologue. *PLoS ONE*, 2013. 8(7): p. e67843.
[71] Liepinsh, E., et al., The death-domain fold of the ASC PYRIN domain, presenting a basis for PYRIN/PYRIN recognition. *J. Mol. Biol.,* 2003. 332(5): p. 1155-63.
[72] Natarajan, A., R. Ghose, and J. M. Hill, Structure and Dynamics of ASC2, a Pyrin Domain-only Protein That Regulates Inflammatory Signaling. *Journal of Biological Chemistry*, 2006. 281(42): p. 31863-31875.
[73] de Alba, E., Structure and Interdomain Dynamics of Apoptosis-associated Speck-like Protein Containing a CARD (ASC). *Journal of Biological Chemistry,* 2009. 284(47): p. 32932-32941.
[74] Ayala, I., et al., An efficient protocol for the complete incorporation of methyl-protonated alanine in perdeuterated protein. *Journal of Biomolecular NMR,* 2009. 43(2): p. 111-119.
[75] Park, H., PYRIN domains and their interactions in the apoptosis and inflammation signaling pathway. *Apoptosis,* 2012. 17(12): p. 1247-1257.
[76] Valmiki, M. and J. Ramos, Death Effector Domain-Containing Proteins. *Cellular and Molecular Life Sciences*, 2009. 66(5): p. 814-830.
[77] Fiory, F., et al., Frontiers: PED/PEA-15, a multifunctional protein controlling cell survival and glucose metabolism. *American Journal of Physiology - Endocrinology And Metabolism,* 2009. 297(3): p. E592-E601.
[78] Danziger, N., et al., Cellular Expression, Developmental Regulation, and Phylogenic Conservation of PEA-15, the Astrocytic Major Phosphoprotein and Protein Kinase C Substrate. *Journal of Neurochemistry,* 1995. 64(3): p. 1016-1025.

[79] Estellés, A., et al., The Major Astrocytic Phosphoprotein PEA-15 Is Encoded by Two mRNAs Conserved on Their Full Length in Mouse and Human. *Journal of Biological Chemistry,* 1996. 271(25): p. 14800-14806.

[80] Renganathan, H., et al., Phosphorylation of PEA-15 switches its binding specificity from ERK/MAPK to FADD. *Biochem. J.,* 2005. 390(3): p. 729-735.

[81] Krueger, J., et al., Phosphorylation of Phosphoprotein Enriched in Astrocytes (PEA-15) Regulates Extracellular Signal-regulated Kinase-dependent Transcription and Cell Proliferation. *Mol. Biol. Cell,* 2005. 16(8): p. 3552-3561.

[82] Araujo, H., et al., Characterization of PEA-15, a major substrate for protein kinase C in astrocytes. *Journal of Biological Chemistry*, 1993. 268(8): p. 5911-5920.

[83] Kubes, M., et al., Endothelin Induces a Calcium-Dependent Phosphorylation of PEA-15 in Intact Astrocytes: Identification of Ser104 and Ser116 Phosphorylated, Respectively, by Protein Kinase C and Calcium/Calmodulin Kinase II *In Vitro*. *Journal of Neurochemistry,* 1998. 71(3): p. 1307-1314.

[84] Trencia, A., et al., Protein Kinase B/Akt Binds and Phosphorylates PED/PEA-15, Stabilizing Its Antiapoptotic Action. *Mol. Cell. Biol.,* 2003. 23(13): p. 4511-4521.

[85] Kitsberg, D., et al., Knock-Out of the Neural Death Effector Domain Protein PEA-15 Demonstrates That Its Expression Protects Astrocytes from TNFα-Induced Apoptosis. *The Journal of Neuroscience,* 1999. 19(19): p. 8244-8251.

[86] Xiao, C., et al., Tumor Necrosis Factor-related Apoptosis-inducing Ligand-induced Death-inducing Signaling Complex and Its Modulation by c-FLIP and PED/PEA-15 in Glioma Cells. *Journal of Biological Chemistry*, 2002. 277(28): p. 25020-25025.

[87] Condorelli, G., et al., PED/PEA-15: an anti-apoptotic molecule that regulates FAS/TNFR1-induced apoptosis. *Oncogene,* 1999. 18(31): p. 4409-15.

[88] Formstecher, E., et al., PEA-15 Mediates Cytoplasmic Sequestration of ERK MAP Kinase. *Developmental Cell*, 2001. 1(2): p. 239-250.

[89] Callaway, K., et al., The Anti-Apoptotic Protein PEA-15 Is a Tight Binding Inhibitor of ERK1 and ERK2, Which Blocks Docking Interactions at the D-Recruitment Site†. *Biochemistry,* 2007. 46(32): p. 9187-9198.

[90] Condorelli, G., et al., PED/PEA-15 gene controls glucose transport and is overexpressed in type 2 diabetes mellitus. *EMBO J.*, 1998. 17(14): p. 3858-66.

[91] Zhang, Y., et al., Regulation of Expression of Phospholipase D1 and D2 by PEA-15, a Novel Protein That Interacts with Them. *Journal of Biological Chemistry,* 2000. 275(45): p. 35224-35232.

[92] Condorelli, G., et al., Protein Kinase C (PKC)-α Activation Inhibits PKC-ζ and Mediates the Action of PED/PEA-15 on Glucose Transport in the L6 Skeletal Muscle Cells. *Diabetes,* 2001. 50(6): p. 1244-1252.

[93] Vigliotta, G., et al., Overexpression of the ped/pea-15 Gene Causes Diabetes by Impairing Glucose-Stimulated Insulin Secretion in Addition to Insulin Action. *Mol. Cell. Biol.*, 2004. 24(11): p. 5005-5015.

[94] Farina, B., et al., NMR backbone dynamics studies of human PED/PEA-15 outline protein functional sites. *FEBS Journal,* 2010. 277(20): p. 4229-4240.

[95] Callaway, K., M. A. Rainey, and K. N. Dalby, Quantifying ERK2-protein interactions by fluorescence anisotropy: PEA-15 inhibits ERK2 by blocking the binding of DEJL domains. *Biochimica et Biophysica Acta (BBA) - Proteins and Proteomics,* 2005. 1754(1-2): p. 316-323.

[96] Kaoud, T. S., et al., Activated ERK2 Is a Monomer *in Vitro* with or without Divalent Cations and When Complexed to the Cytoplasmic Scaffold PEA-15. *Biochemistry,* 2011. 50(21): p. 4568-4578.

[97] Twomey, E. C., et al., Substantial Conformational Change Mediated by Charge-Triad Residues of the Death Effector Domain in Protein-Protein Interactions. *PLoS ONE,* 2013. 8(12): p. e83421.

[98] Mace, P. D., et al., Structure of ERK2 bound to PEA-15 reveals a mechanism for rapid release of activated MAPK. *Nat. Commun.*, 2013. 4: p. 1681.

[99] Doti, N., et al., Residues 762-801 of PLD1 mediate the interaction with PED/PEA15. *Molecular Biosystems*, 2010. 6(10): p. 2039-2048.

[100] Viparelli, F., et al., Targeting of PED/PEA-15 Molecular Interaction with Phospholipase D1 Enhances Insulin Sensitivity in Skeletal Muscle Cells. *Journal of Biological Chemistry*, 2008. 283(31): p. 21769-21778.

[101] Farina, B., et al., Molecular basis of the PED/PEA15 interaction with the C-terminal fragment of phospholipase D1 revealed by NMR spectroscopy. *Biochim. Biophys. Acta,* 2013. 1834(8): p. 1572-80.

In: Nuclear Magnetic Resonance (NMR) ISBN: 978-1-63117-768-2
Editor: Doddapuneni Krishna Rao © 2014 Nova Science Publishers, Inc.

Chapter 6

THEORIES AND APPLICATIONS IN SOLID-STATE NUCLEAR MAGNETIC RESONANCE SPECTROSCOPY

Eugene Stephane Mananga[1*], Betina Suzanne Mananga[2] and Alicia Errolene Reid[3]

[1]Harvard University, Harvard Medical School, and Massachusetts General Hospital, Center for Advanced Medical Imaging Sciences, Division of Nuclear Medicine and Molecular Imaging Physics, Department of Radiology, Boston, MA, US

[2]Université Paris-Sud, Faculté des Sciences d'Orsay, Orsay, Cedex, France

[3]The City University of New York, Medgar Evers College, Brooklyn, NY, US

ABSTRACT

Recent advances in the theory of solid state nuclear magnetic resonance (NMR) such as Floquet-Magnus expansion and Fer expansion, address alternative methods for solving a time-dependent linear differential equation which is a central problem in solid-state NMR. The power and the salient features of these theoretical approaches that are helpful in describing the time evolution of the spin system at all times are

---

* Corresponding author: Telephone +1 646 345 4613; Fax: + 1 617 726 6165; E-mail address: emananga@pet.mgh.harvard.edu (Eugene S. Mananga).

sketch out within. This chapter presents a broad view of the algorithms of spin dynamics, based on successful and useful theories (average Hamiltonian theory, Floquet theory, Floquet-Magnus expansion, Fer expansion). These algorithms provide procedures to control and describe spin dynamics in solid-state NMR. Applications of these theoretical methods to stroboscopic and synchronized manipulations, non-synchronized experiments, multiple incommensurate frequencies, magic-angle spinning samples, are illustrated. We also reviewed the theory of solid-state NMR spectroscopy, with specific discussions of magnetic interactions in the solid state such as dipole-dipole, chemical-shift and quadrupolar.

I. INTRODUCTION

Spectroscopic methods, which study the interaction of energy with matter, have been used extensively and broadly in fields ranging from quantum physics, chemistry and biology to astronomy. Of the different spectroscopic methods, nuclear magnetic resonance (NMR) has proven an invaluable technique for deriving structural information. Nuclear magnetic resonance is now established as a valuable analytical tool for applications in fields varying from chemistry, molecular biology, and physics to medical imaging. Based on the interaction of the magnetic moment of a nucleus and an external magnetic field, structural and dynamic information may be derived from NMR spectra for both samples in the liquid state and solid state [1-5].

Liquid/solution-state NMR, based on the relative ease to obtain high resolution spectra, is a much more mature tool for chemical characterization as compared to solid-state NMR. Nonetheless, for many applications as the analysis of membrane proteins [6-10], polymers [11-15], glasses [16, 17] and battery materials [11-15], it is necessary for samples to be in the solid state. To understand why lower resolution spectra are typically derived for samples in the solid state as compared to the liquid state one should consider the NMR Hamiltonian. The full NMR Hamiltonian may be given by [10, 18-20]:

$$H = H_Z + H_{RF} + H_J + H_{CS} + H_{DD} + H_Q,$$

where Z = Zeeman splitting; RF = radio frequency of applied electromagnetic radiation; J = J coupling; CS = chemical shift coupling; D = dipole-dipole coupling and Q = quadrupolar coupling. In liquid-state NMR the relatively sharp/well resolved signals obtained are a result of the random molecular reorientations taking place in the liquid state which cause an averaging out of the anisotropic contributions to the chemical shift and the dipolar interactions. Though well resolved spectra are obtained from the averaging out of the dipolar and anisotropic contributions, this is accompanied by a loss of dynamic information. In the solid state, broad peaks are typically obtained because such molecular reorientations do not take place; however, on the positive side much valuable information is conserved.

The challenge in solid-state NMR (SSNMR) then becomes, how do we preserve structural and dynamic information while still deriving well resolved and useful spectra. A number of techniques have been developed to manipulate the Hamilitonian such that high resolution spectra are obtained while maintaining sufficient structural information.

These techniques include magic-angle spinning [21-24], cross polarization [26, 27] and the use of special pulse sequences [28-30]. In addition to these clever techniques used for improved resolution, further advancements are still much needed for the greater utility of SSNMR. Advances in the field of SSNMR will be dependent on improvements in hardware (such as stronger magnets) [31]; in computational tools (such as simulation software) [32] and most importantly in the theoretical tools used to design and analyze new SSNMR experiments [31].

In this chapter we will look in detail at the established and emerging theoretical methods used in the design and evaluation of pulse sequences for manipulation of the Hamiltonian in SSNMR experiments. The average Hamilitonian theory (AHT) [33, 34], the Floquet theory (FLT) [35-38], the Fer expansion (FE) [39, 40] and the Floquet-Magnus expansion (FME) [41, 42] along with the advantages and limitations of each of these theoretical approaches will be discussed. The average Hamilitonian theory and the Floquet theory are the most widely applied theories for qualitative problems in SSNMR. AHT, introduced by Haeberlen and Waugh to SSNMR in the late sixties [33], is used to describe spin interactions by considering the average of the effective Hamiltonian after the RF pulse. Unlike the AHT, the FLT introduced in the eighties describes the time evolution of the spin system at all times after the RF pulse. The Fer expansion (FE) offers an alternative expansion scheme to the Magnus expansion used in the AHT, but is still in its infancy as compared to AHT and FLT. The FE was illustrated in NMR by considering simple applications such as the calculation of Bloch-Siegert shift in NMR and the analysis of the heteronuclear decoupling in solid-state NMR [39]. The newest of the theoretical approaches, the Floquet-Magnus expansion

(FME) [41], is a unique approach in spin dynamics that shed new lights on AHT and FLT. This scheme was recently introduced to solid-state NMR and applied to many NMR interactions [41, 43-46].

II. Average Hamiltonian Theory

The average Hamiltonian theory has evolved as a powerful technique in the development of high resolution NMR spectroscopy. The Magnus expansion [42, 47-60] has been systematically used in NMR, in particular in solid-state NMR where it forms the basis of AHT. This expansion was first applied to NMR in 1968 by Evans [47], Haeberlen and Waugh [33]. Since its introduction, the ME has been instrumental in the development of improved techniques in NMR spectroscopy [42, 48]. Developed by John Waugh and co-workers in 1968, the AHT approach became the main tool to study the dynamics of spin systems subject to an RF perturbation. This approach is the most commonly used method to treat theoretical problems in solid-state NMR and have been used sometimes abusively [61]. A basic understanding of the AHT involves considering a time dependent Hamiltonian governing the spin system evolution, and describing the effective evolution by an average Hamiltonian $\overline{H}$ within a periodic time (T). This is satisfied only if is periodic (T) and the observation is stroboscopic and synchronized with period (T). The AHT has been widely used in the NMR literature in the development of multiple pulse sequences and in the context of both decoupling and recoupling experiments. This technique allows for stroboscopic manipulation of spins and spin interactions by radio-frequency pulses. By considering the average or effective Hamiltonian of the RF pulse train, the AHT also explains how periodic pulses can be used to transform the symmetry of selected interactions in coupled and many-spin systems.

II.1. Useful Equations and Propagators

The main result of the AHT is given by [33, 34, 62-64]

$$\overline{H}_0 = \overline{H}_0^{(0)} + \overline{H}_0^{1} + \overline{H}_0^{2} + \ldots \tag{1}$$

with

$$\overline{H}_0^{(0)} = \frac{1}{t_c}\int_0^{t_c} dt_1 \widetilde{H}_0(t_1), \tag{2}$$

$$\overline{H}_0^{(1)} = \frac{-i}{2t_c}\int_0^{t_c} dt_2 \int_0^{t_2} dt_1 \left[\widetilde{H}_0(t_2), \widetilde{H}_0(t_1)\right], \tag{3}$$

where is the toggling frame Hamiltonian, which is the Hamiltonian in the time-dependent interaction.

The evolution propagator is obtained by expressing the average Hamiltonian and using the Magnus expansion. The AHT/evolution propagator is given by:

$$U(t_c) = \exp\left\{-i\overline{H}(t_c)t_c\right\} \tag{4}$$

where is the average Hamiltonian and corresponds to the period of a periodic Hamiltonian $H(t)$.

II.2. Applications

The AHT has been gradually applied to many theoretical problems in solid-state NMR [61]. We recently used this technique to compare the efficiency of the magic echo cycle and the simple two-pulse quadrupolar echo sequence [62, 63, 65-67]. Outline below are some selected examples for the application of AHT in the development and analysis of pulse sequences.

II.2.1. Symmetry in the Design of NMR Multiple-Pulse Sequences Using AHT

The average Hamiltonian theory was developed for a class of symmetrical radio-frequency pulse sequences used for NMR of rotating solids [68]. Theorems are presented which allow one to predict the elimination of many average Hamiltonian terms. These findings are used to study the problem of heteronuclear decoupling in the presence of rapid magic angle spinning. Sequences which minimize the number of heteronuclear terms at the same time as recoupling the homonuclear interactions of the irradiated spins are presented. The heteronuclear decoupling performance of the new sequences is significantly better than that obtained with an unmodulated radio-frequency field. The decoupling performance is improved further by breaking the pulse sequence symmetry in a controlled fashion [68]. The problem of heteronuclear decoupling requires the application of the AHT up to at least the second order. Several groups have developed symmetry theorems for suppressing average Hamiltonian terms in the presence of time- independent internal spin interactions, but most of these results do not apply strictly in a rotating sample. The average Hamiltonian symmetries may also be useful for a variety of purposes in the NMR of rotating solids, such as homonuclear and heteronuclear recoupling, higher order multiple-quantum excitation, and the selection of isotropic interactions. For instance, AHT was used in the calculation of: homonuclear dipole-dipole couplings and chemical shifts for the simple pulse sequence (delta-function) called DRAMA [69], continuous rf irradiation with the pulse sequence called 2Q-HORROR [70], finite-pulse recoupling sequences such as finite-pulse Radio-Frequency-Driven Recoupling (fpRFDR) [71, 72], chemical-shift-driven recoupling including rotational resonance [73-80] and radio-frequency-driven-recoupling (SEDRA) [73-79]. Furthermore, the symmetry rules may readily be generalized to the case of simultaneous rf sequences on several irradiation channels. This may have relevance for cross-polarization, and for the problem of establishing good heteronuclear decoupling during homonuclear recoupling sequences. Malcolm Levitt and his collaborators have developed an approach to the development of recoupling sequence that relies on the general symmetry properties of pulse sequences. This approach produces *selection rules* that reveal which types of interactions can be recoupled by a sequence with a given symmetry. First and second class sequences corresponding to "C" (C7 and POST-C7 are good examples) and "R" sequences belonging to two distinct symmetry classes have been described [45, 81-86].

II.2.2. Composite Pulses Sequences

Composite pulses are the decomposition of a single pulse into the product of a number of constituent pulses [87-89]. These schemes have been developed to compensate for rf inhomogeneity, off-resonance effects, and phase distorsion. These pulse sequences have been found to be useful in a number of experimental applications where it is necessary to overcome instrumental shortcomings. One useful application of composite pulses is the effective creation of unitary operators which are not easily accessible. The average Hamiltonian theory has been very useful in designing and constructing composite pulses sequences. For example, theoretical precedures for constructing the composite pulses that compensate for the inhomogeneities of the electric field gradient and radio frequency field strength for nuclear quadrupole resonance spectroscopy uses a technique similar to the AHT [90].

II.3. Advantages and Limitations

The AHT is not applicable to Hamiltonians with multiple basic frequencies [91, 92]. No single cycle time can be defined if the two frequencies are incommensurate: magic angle spinning (MAS) and radiofrequency irradiation must be synchronized or time-scale separated, multiple irradiations must be synchronized or time-scale separated. Only stroboscopic observation is allowed after a full cycle time. Nyquist frequency equals the basic frequency while side bands appear at integer multiples of the basic frequency. Sidebands in MAS spectra cannot be described and are folded back onto the center band. Sidebands in rf irradiation schemes are neglected and folded back onto the center band. Convergence of the series expansion of the Hamiltonian can also be a problem. Usually the basic frequency has to be larger than the transition frequencies in the Hamiltonian.

Though holding well for static experiments, the AHT suffers from the following shortcomings [34, 91-97]:

a) This technique does not sufficiently describe the case of magic angle spinning (MAS) spectra. In the case of MAS, the signal is usually observed continuously with a time resolution much shorter than the rotor period.
b) One has to be able to define a single basic frequency as well as a cycle time of the Hamiltonian.
c) The AHT cannot be used with multiple incommensurate time-dependent processes in solid-state NMR such as sample rotation and non-synchronized radio frequency irradiation.
d) The convergence of the series expansion of the Hamiltonian can be a problem and the basic frequency has to be larger than the transition frequencies in the Hamiltonian.

Recently, the validity of the AHT method was probed for quadrupolar nuclei [64]. The investigation showed that the AHT method was less efficient to predict the dynamics of the spin system as the quadrupolar spin nuclei dimension increased. This was attributed to the Hilbert space becoming very large and leading to the contribution of non-negligible higher order terms in the Magnus expansion being truncated. For instance, considering a simple two-pulse sequence for refocusing the quadrupolar Hamiltonian, shown in Figure 2, Mananga et al. showed that the ability of the AHT to predict the spin dynamics depends on the size of the spin system [64].

III. Floquet Theory

The FLT, introduced to the NMR community in the early 1980's simultaneously by Vega [38] and Maricq [37], is another powerful approach that offers a way to describe the time evolution of the spin system at all times and is able to handle multiple incommensurate frequencies. Floquet theory is an exact method and does not imply any assumptions or approximations. This theory provides a more general approach to the AHT and is useful in discussing the convergence of the expansion. The theory maps the finite-dimensional time-dependent Hilbert space onto an infinite-dimensional but time-independent Floquet space [98-113]. The Floquet description requires an additional Fourier space to describe the quantization of the motional process. Matrix-based Floquet description leads to a correct description of time-dependent Hamiltonians including the side bands. The FLT approach allows the computation of the full spinning sideband pattern that is of importance in many MAS experimental circumstances for obtaining information on anisotropic sample properties. The FLT has been applied satisfactorily to simple-spin systems, spin-pair systems to study important NMR phenomena including rotational-resonance, composite pulse sequence designing, field-dependent chemical shifts, cross-polarization dynamics, two-dimensional solution NMR experiments, and the dynamic characterization of exchanging spin systems.

The general description of the FLT is equally applicable to dipolar systems as well as to quadrupolar nuclear spin systems. However, spin systems with large quadrupolar couplings may violate the convergence conditions for the expansions employed to evaluate the Floquet matrices. An important question to answer is to what extent can the FLT be used in NMR without loss of its conceptual framework? Probing the validity of the FLT for quadrupolar nuclei, including those with spin I=1, 3/2, 5/2, and 7/2, by analyzing a simple pulse sequence can be beneficial to the NMR community

[64]. While the FLT scheme provides a more universal approach for the description of the full time dependence of the response of a periodically time-dependent system, it is most of the time impractical. Analytical calculations are limited to small spin systems and it is difficult to get physical insight from the matrix representation [114]. For instance, Matti Maricq obtained results that showed the Floquet theory and the average Hamiltonian theory are equal for each of the first two orders but comparison of higher orders is more difficult [34,37].

The full Floquet Hamiltonian has an infinite dimension and it is often not very easy to understand its implications on the time evolution of the spin system. Matrices for multi-mode Floquet calculations can become intractable. Massive reduction in dimensionality by truncation of the Fourier dimensions can introduce artifacts. In the literature, problems with up to three frequencies have been treated, but the demand for experiments that require four frequencies for a full description is increasing. For example, non-cyclic multiple-pulse sequences like two-pulse phase-modulated (TPPM) decoupling experiments acquire four frequencies under double rotation (DOR) and there are some other obvious problems with four frequencies like triple-resonance CW radio frequency irradiation under MAS. To the best of the authors' knowledge, the Floquet theory has not been used to analyze solid-state NMR Hamiltonians with four or more frequencies.

III.1. Useful Equations and Propagators

III.1.1. Operator-Based Floquet Theory [91, 92, 114]

Operator-based Floquet theory in solid-state NMR gives the following results.

Single-mode Floquet Hamiltonian: we assume a periodic time-dependent Hamiltonian that can be expanded in a Fourier series of the form

$$\hat{H}(t) = \sum_{n=-\infty}^{\infty} {}^{(n)}\hat{H} e^{in\omega_m t} \tag{5}$$

where ω_m is the characteristic frequency and the operators can be viewed as Fourier coefficients of the Hamiltonian with respect to this frequency. The effective Hamiltonian for a single-mode Floquet Hamiltonian can be obtained to different orders given by

$$\overline{\hat{H}} = {}^{(0)}\hat{H} - \frac{1}{2}\sum_{n\neq 0}\frac{\left[{}^{(-n)}\hat{H}, {}^{(n)}\hat{H}\right]}{n\omega_m} + \frac{1}{2}\sum_{n\neq 0}\frac{\left[\left[{}^{(n)}\hat{H}, {}^{(0)}\hat{H}\right], {}^{(-n)}\hat{H}\right]}{(n\omega_m)^2} + \frac{1}{3}\sum_{k,n\neq 0}\frac{\left[{}^{(n)}\hat{H}, \left[{}^{(k)}\hat{H}, {}^{(-n-k)}\hat{H}\right]\right]}{kn\omega_m^2} + \ldots \tag{6}$$

Bimodal Floquet Hamiltonian: we assume a periodic time-dependent Hamiltonian that can be expanded in a Fourier series of the form

$$\hat{H}(t) = \sum_{n=-\infty}^{\infty}\sum_{k=-\infty}^{\infty} {}^{(n,k)}H e^{in\omega_r t} e^{ik\omega_m t} \tag{7}$$

where ω_m and are the characteristic frequencies and the operators can be viewed as Fourier coefficients of the Hamiltonian with respect to these frequencies. The effective Hamiltonian for a bimodal Floquet Hamiltonian can be obtained to different orders given by

$$\overline{\hat{H}} = \sum_{n_0,k_0} {}^{(n_0,k_0)}\hat{H} - \sum_{n_0,k_0}\frac{1}{2}\sum_{\nu,\kappa}\frac{\left[{}^{(n_0-\nu,k_0-\kappa)}\hat{H}, {}^{(\nu,\kappa)}\hat{H}\right]}{\nu\omega_r + \kappa\omega_m} + \ldots \tag{8}$$

with

$$n_0\omega_r + k_0\omega_m = 0 \tag{9}$$

and

$$\nu\omega_r + \kappa\omega_m \neq 0 \tag{10}$$

For bimodal Floquet problem, resonance conditions will always involve both frequencies.

Triple-mode Floquet Hamiltonian: we assume a periodic time-dependent Hamiltonian that can be expanded in a Fourier series of the form

$$\hat{H}(t) = \sum_{n=-\infty}^{\infty}\sum_{k=-\infty}^{\infty}\sum_{l=-\infty}^{\infty} {}^{(n,k,l)}He^{in\omega_r t}e^{ik\omega_m t}e^{il\omega_p t} \tag{11}$$

where ω_m, ω_r, and are the characteristic frequencies and the operators can be viewed as Fourier coefficients of the Hamiltonian with respect to these frequencies. The effective time-independent Hamiltonian for a triple-mode Floquet Hamiltonian can be obtained to different orders given by

$$\overline{\hat{H}} = \sum_{n_0,k_0,l_0} {}^{(n_0,k_0,l_0)}\hat{H} - \sum_{n_0,k_0,l_0} \frac{1}{2} \sum_{\nu,\kappa,\lambda} \frac{\left[{}^{(n_0-\nu,k_0-\kappa,l_0-\lambda)}\hat{H}, {}^{(\nu,\kappa,\lambda)}\hat{H} \right]}{\nu\omega_r + \kappa\omega_m + \lambda\omega_p} + \ldots \qquad (12)$$

with

$$n_0\omega_r + k_0\omega_m + l_0\omega_p = 0 \qquad (13)$$

and

$$\nu\omega_r + \kappa\omega_m + \lambda\omega_p \neq 0 \qquad (14)$$

For triple-mode and higher Floquet problems, resonance conditions involving all frequencies are possible but also resonance conditions that involve only a subset of the frequencies are also possible. Such resonance conditions reveal problems that are partially resonant and partially non-resonant, depending on the frequencies we look at.

III.1.2. Matrix-Based Floquet Theory

The Floquet Hamiltonian, the density operator and the detector operators are extensively presented in the literature [91, 92, 114]. In a flow chart summarizing the main steps in the application of Floquet theory in NMR, M. Leskes et al. described the time dependent periodic Hamiltonian, the Fourier representation, the Floquet representation (full diagonalisation, Floquet propagator, density matrix, and observables) and the sequence design (Block diagonalisation, evaluation of the NMR experiment, and single mode) [91].

The FLT propagator is given by:

$$U(t) = P(t).\exp\{-iH_F t\} \qquad (15)$$

where the operator is periodic with the period t_c, while the Floquet Hamiltonian is time-independent.

III.2. Applications

Several MAS NMR experiments on spin systems with a periodically time-dependent Hamiltonian were extensively discussed in the recent articles by Leskes et al. [91] and Scholz et al. [92]. For many NMR experiments, understanding the spin dynamics requires a wise choice of the interaction frame in which the Hamiltonian is presented. A transformation of the Hamiltonian to such a frame often leads to periodic time dependences which can be removed by an additional transformation to the Floquet representation. After deriving the Floquet Hamiltonian, the van Vleck transformation was applied to obtain an effective Hamiltonian. Leskes et al. [91] described all examples using the Floquet operators introduced by Boender and coworkers [115]. The authors Leskes and Scholz applied the Floquet theory for the investigation of numerous cases as: MAS, rotational-resonance recoupling, continuous wave (CW) irradiation on a single spin species, DARR and MIRROR recoupling, simultaneous CW irradiation on two different spin species, phase-alternating (XiX) irradiation on a single spin species, CW irradiation on one and XiX irradiation on a second spin species, phase-modulated Lee-Goldburg decoupling, C-type and R-type sequences, TPPM decoupling, chemical shift anisotropy (CSA) spectra during MAS experiments, recoupling under MAS (rotational resonance, recoupling and decoupling with CW irradiation), heteronuclear decoupling, cross polarization, homonuclear decoupling, quadrupolar nuclei and dynamic MAS [91, 92]. Below we present the cases of CW decoupled under MAS and the multipole-multimode Floquet theory (MMFT) in NMR.

III.2.1. CW Decoupled under MAS

We assume a two-spin system, I-S with rf irradiation on the I spin. The Hamiltonian describing this system, in the rotating frame, is giving by

$$H(t) = \sum_{n=-2}^{2} \omega_S^{(n)} e^{in\omega_r t} S_Z - \sum_{n=-2}^{2} \omega_I^{(n)} e^{in\omega_r t} I_X - \sum_{\substack{n=-2 \\ n\neq 0}}^{2} \omega_{SI}^{(n)} e^{in\omega_r t} 2S_Z I_X + \omega_1 I_{IZ} \tag{16}$$

The coordinate system of the I spin is tilted by 90^0 around the y axis such that the rf is along the z axis. The transformation into an interaction frame with the rf:

$$\widetilde{H}(t) = e^{(i\omega_1 I_Z t)} H(t) e^{(-i\omega_1 I_Z t)} \tag{17}$$

gives

$$\widetilde{H}(t) = \sum_{n=-2}^{2} \omega_S^{(n)} e^{(in\omega_r t)} S_Z - \sum_{n=-2}^{2} \omega_I^{(n)} e^{(in\omega_r t)} (\frac{1}{2} I^+ e^{i\omega_1 t} + \frac{1}{2} I^- e^{(-i\omega_1 t)})$$
$$- \sum_{\substack{n=2 \\ n\neq 0}}^{2} \omega_{SI}^{(n)} e^{(in\omega_r t)} 2S_Z (\frac{1}{2} I^+ e^{i\omega_1 t} + \frac{1}{2} I^- e^{(-i\omega_1 t)}) \tag{18}$$

The Fourier coefficients of the above Hamiltonian $\widetilde{H}(t)$ can be written:

$$^{(0,0)}\widetilde{H} = \omega_S^{(0)} S_Z \tag{19}$$

$$^{(n,0)}\widetilde{H} = \omega_S^{(n)} S_Z \tag{20}$$

$$^{(0,\pm 1)}\widetilde{H} = -\frac{\omega_I^{(0)}}{2} I^{\pm} - \omega_{SI}^{(0)} S_Z I^{\pm} \tag{21}$$

$$^{(n,\pm 1)}\widetilde{H} = -\frac{\omega_I^{(n)}}{2} I^{\pm} - \omega_{SI}^{(n)} S_Z I^{\pm} \tag{22}$$

$$^{(n,\pm 2)}\widetilde{H} = 0 \tag{23}$$

The first index describes the rotation in real space (MAS) while the second index describes the rotation in spin space (interaction frame).

Decoupling is irradiation outside any resonance condition, i.e., only and have to be considered. The effective Hamiltonian is reduced to:

$$\overline{\hat{H}} = {}^{(0,0)}\hat{H} - \frac{1}{2}\sum_{\nu,\kappa} \frac{\left[{}^{(-\nu,-\kappa)}\hat{H}, {}^{(\nu,\kappa)}\hat{H}\right]}{\nu\omega_r + \kappa\omega_m} + \ldots$$

$$= \omega_S^{(0)} S_Z$$

$$+ \frac{1}{4}\sum_{\nu}\left(\frac{\omega_I^{(\nu)}\omega_I^{(-\nu)} + \omega_{SI}^{(\nu)}\omega_{SI}^{(-\nu)}}{\nu\omega_r + \omega_1} I_Z + \frac{\omega_I^{(\nu)}\omega_{SI}^{(-\nu)} + \omega_{SI}^{(\nu)}\omega_I^{(-\nu)}}{\nu\omega_r + \omega_1} 2S_Z I_Z \right)$$

$$- \frac{1}{4}\sum_{\nu}\left(\frac{\omega_I^{(\nu)}\omega_I^{(-\nu)} + \omega_{SI}^{(\nu)}\omega_{SI}^{(-\nu)}}{\nu\omega_r - \omega_1} I_Z + \frac{\omega_I^{(\nu)}\omega_{SI}^{(-\nu)} + \omega_{SI}^{(\nu)}\omega_I^{(-\nu)}}{\nu\omega_r - \omega_1} 2S_Z I_Z \right) \quad (24)$$

Such a result cannot be obtained with average Hamiltonian theory without making assumptions about the ratio of ω_1 and ω_r.

III.2.2. Multipole-Multimode Floquet Theory in NMR

The Multipole theory proposed by Sanctuary [116-118] to describe spin dynamics in NMR differs from the conventional Floquet theory in that the dynamics is treated in the operator space (Liouville space) with an explicit time-dependent density operator. While the Sanctuary method has been successful for describing the dynamical behavior, it is often limited to calculations involving only the experimentally detectable magnetization. In addition, it fails to provide insights into the multiple-quantum NMR phenomena and polarization transfer experiments that involve relaxation. Despite the spin Hamiltonian being described in terms of irreducible tensor operators, it is not yet common practice to express the spin basis in terms of operators which are also irreducible under rotations [119-121]. This method has been exploited extensively in the multipole formalism, where the basis operators are deliberately chosen to be invariant under the rotation group SO(3), thereby extending their range of applicability from single spins to multiple spins. In the multipole approach, we use the following compact notation employed by Sanctuary et al. [122] for the density operator

$$\rho(t) = \frac{1}{\prod_{i=1}^{N}(2I_i + 1)} \sum_{k,q,\alpha} \Phi_q^{(k)}(\overline{\alpha}, t) T^{(k)q}(\overline{\alpha}) \quad (25)$$

where is representative of all the quantum numbers illustrated previously for the description in the operator space [116]. In a similar vein, the Hamiltonian (assumed to be time independent) of the system is also expressed in the same basis as that of the density operator and is represented by

$$H = \sum_{l,m,\overline{\alpha}} C_{lm(\overline{\alpha})} T^{(l)m}(\overline{\alpha}) \tag{26}$$

where and denote the rank and component of the tensor, respectively, with being the standard case in NMR experiments. The interaction parameters corresponding to the Zeeman, dipole-dipole, and quadrupolar interactions are represented by the coefficients. We have set in this study.

Substituting $\rho(t)$ and H in the following Liouville equation

$$i\frac{d\rho(t)}{dt} = [H, \rho(t)] \tag{27}$$

we obtain a set of differential equations spanning the entire operator space:

$$i\frac{d\Phi_q^{(k)}(\overline{\alpha},t)}{dt} = \sum_{k',q',\overline{\alpha}'} \sum_{l,m,\alpha_1} C_{lm}(\overline{\alpha_1}) Tr\left\{T^{(k)-q}(\overline{\alpha}) \times L^{(l)m}(\overline{\alpha_1}) T^{(k')q'}(\overline{\alpha}')\right\} \Phi_{q'}^{(k')}(\overline{\alpha}',t) \tag{28}$$

where $L^{lm}(\overline{\alpha_1}) = \left[T^{lm}(\overline{\alpha_1})\right]$, corresponds to the Liouville super-operator. The above equation can be conveniently re-expressed in matrix form as

$$i\frac{d\Phi_q^{(k)}(\overline{\alpha},t)}{dt} = [A]_{kq,k'q'} \Phi_{q'}^{(k')}(\overline{\alpha}',t) \tag{29}$$

where the spin polarizations $\Phi_{q'}^{(k')}(\overline{\alpha}',t)$ are represented by column vectors with [A] representing the super-matrix whose elements are evaluated using the Wigner-Eckart theorem [123,124]. Details for the evaluation of the super-matrix can be found in the literature [116].

Now we assume the Hamiltonians to be explicitly time dependent (MAS, rf) and we solve the above set of differential equations involving time-dependent coefficients. Such periodic time-dependent phenomena are easily analyzed using Floquet theory. However, existing bases employed in the theoretical treatment of the Floquet theory are often difficult to extend to multiple spins and systems. To this end, the MMFT approach is a viable alternative scheme for studying the dynamics under synchronous and asynchronous time modulations in solid-state NMR experiments. The time-dependent spin Hamiltonian, the density operator, and the Liouville super-operator in a Fourier series can be written as [116]

$$H(t) = \sum_{n_{1\to m}=-\infty}^{\infty} H_{n_{1\to m}} e^{it\omega.n} \tag{30}$$

$$\rho(t) = \sum_{n_{1\to m}=-\infty}^{\infty} \sum_{k,q,\underline{\alpha}} \Phi^{(k)}_{q,n_{1\to m}}(\overline{\alpha},t) T^{(k)q}(\overline{\alpha}) e^{it\omega.n} \tag{31}$$

and

$$L(t) = \sum_{n_{1\to m}=-\infty}^{\infty} L(\overline{\alpha_1})^{(l)m}_{n_{1\to m}} e^{it\omega.n} \tag{32}$$

where the following notation is used: $n_{1\to m} = n_1, n_2, \ldots, n_m$, $A_{n_{1\to m}} = A_{n_1, n_2, \ldots n_m}$, $\omega = \{\omega_1, \omega_2, \ldots, \omega_m\}$, $n = \{n_1, n_2, \ldots, n_m\}$, and $\omega.n = \omega_1 n_1 + \omega_2 n_2 + \ldots + \omega_m n_m$. represents the Fourier index and ω_i represent the frequency associated with a particular time modulation. Substituting the Fourier expansions of the density operator and the Liouville super-operator in the Liouville equation, the following new set of coupled differential equations spanning an infinite dimensional vector space, with time-independent coefficients are obtained:

$$i\frac{d\Phi(\overline{\alpha},t)^{(k)}_{q,n_{1\to m}}}{dt} = \sum_{n'_{1\to m}=-\infty}^{\infty} \sum_{k',q',\underline{\alpha}'} \sum_{l,m,\underline{\alpha_1}} \left(Tr\left\{T^{(k)-q}(\overline{\alpha}) \times L(\overline{\alpha_1})^{lm}_{n_{1\to m}-n'_{1\to m}} T^{(k')q'}(\overline{\alpha'})\right\} + (n.\omega)\delta\right)\Phi(\overline{\alpha'},t)^{(k')}_{q',n'_{1\to m}} \tag{33}$$

The notation $L(\overline{\alpha_1})^{(l)m}_{n_{1\to m}-n'_{1\to m}}$ includes the interaction coefficients as well as the spin and Fourier operators. The above equation can be re-expressed in matrix notation,

$$i\frac{d\Phi^{(k)}_{q,n_{1\to m}}(\overline{\alpha},t)}{dt}=\left[\ \left[A\right]_{kq,k'q'}\ \right]_{n_{1\to m},n'_{1\to m}}\Phi^{(k')}_{q',n'_{1\to m}}(\overline{\alpha'},t) \tag{34}$$

where $\left[\ \left[A\right]_{kq,k'q'}\ \right]_{n_{1\to m},n'_{1\to m}}$ represents the super-matrix defined in the infinite dimensional Floquet-Liouville space. Subsequently, the Floquet density operator and the Hamiltonian operator may be represented by

$$\rho_F(t)=\sum_{n_{1\to m}=-\infty}^{\infty}\sum_{k,q,\underline{\alpha}}\Phi^{(k)}_{q,n_{1\to m}}(\overline{\alpha},t)T^{(k)q}(\overline{\alpha})F^1_{r_1}F^2_{r_2}...F^m_{r_m} \tag{35}$$

$$H_F=\sum_{r_{1\to m}=-\infty}^{\infty}\sum_{k,q,\underline{\alpha}}T^{(k)q}(\overline{\alpha})F^1_{r_1}F^2_{r_2}...F^m_{r_m}+\sum_{i=1}^{m}\omega_i N^i \tag{36}$$

The Floquet Hamiltonian is represented using an operator basis constructed by the direct product of operators defined both in the spin ($T^{(k)q}$) as well as the Fourier dimensions ($F^m_{r_m}$, corresponding to m^{th} time modulation) with the off-diagonality represented by the indices q and respectively. This approach used here provides analytical insights in spite of the infinite dimensionality of the problem. This can be validated by describing an analytical solution in the form of effective Hamiltonians obtained via the contact or van Vleck transformation procedure [116].

III.3. Advantages and Limitations

The Floquet description needs an additional Fourier space to describe the quantization of the motional process. The Floquet theory is an exact method and does not imply any assumptions or approximations. The expressions of the effective Hamiltonian in the spin-Hilbert space without detailed knowledge of the structure of the spin Hamiltonian can be calculated. Results of the operator-based Floquet theory are given above in the equations in subsection (III.1.2.).

The operator-based Floquet theory can easily be expanded to multiple incommensurate frequencies with expressions that are independent of the detailed structure of the spin-Hilbert-space Hamiltonian leading to an effective Hamiltonian that can be written as an analytical operator expression. The operator-based Floquet theory can be used to treat systems with many spins as long as the commutators can be calculated. However, this operator cannot treat resonance conditions that involve simultaneous transition in Fourier space and spin space.

The matrix-based Floquet description leads to an accurate description of time-dependent Hamiltonians including the sidebands and they can be combined with perturbation theory to get insight into processes that involve combined Fourier-space and spin-space resonance conditions. It is very difficult to get physical insight from the matrix representation and analytical calculations are limited to small spin systems. For multi-mode Floquet calculations, matrices can become very large, hence require truncation of the Fourier dimensions which introduce artifacts. In the numerical simulations, time slicing of the time-dependent Hamiltonian is almost always faster and simpler to implement.

IV. Fer Expansion

The Fer expansion was introduced recently to the NMR community by Madhu and Kurur [39] via the effect of Bloch-Siegert shift and heteronuclear dipolar decoupling [125, 126]. This new approach in the field of NMR was formulated more than half a century ago by Fer [40]. This theory is another alternative expansion scheme for solving the time-dependent Schrodinger differential equation. Indeed, from the point of view of physical applications, the Magnus expansion has been extensively used in a variety of cases while, until recently, the Fer expansion has been either ignored or misquoted [51]. This FE approach is still in its infancy and can be considered to be complimentary to the Magnus expansion (AHT). While the efficiency of Fer expansion seems obvious, more work is required for the scheme to overcome difficulties involving cases as non-periodic and non-cyclic cases.

IV.1. Useful Equations and Propagators

The formalism of Fer expansion expresses the solution to the differential time-dependent Schrodinger equation in the form of an infinite-product of series of exponentials as [39]

$$U(t) = \prod_{k=1}^{\infty} e^{F_k(t)} = e^{F_1(t)} e^{F_2(t)} ... \tag{37}$$

A knowledge of the Hamiltonian allow an easy evaluation of

$$F_1 = \int_0^t dt' H(t') \tag{38}$$

and

$$H_F^1 = e^{-F_1(t)} H e^{F_1(t)} - \int_0^1 dx e^{-xF_1(t)} H e^{xF_1(t)} \tag{39}$$

The following iterative formula were derived

$$F_n = -i^n \int_0^t dt' H_F^{(n-1)}(t') \tag{40}$$

$$H_F^{(0)} = H, \tag{41}$$

$$H_F^{(n)} = -\frac{1}{2}\left[F_n, H_F^{(n-1)}\right] + \frac{1}{3}\left[F_n, \left[F_n, H_F^{(n-1)}\right]\right] + ... \tag{42}$$

where. Therefore, the Fer expansion involves a series of nested commutators resulting in $H_F^{(n)}$. The Fer expansion differs from the Magnus approach in the form of the correction terms.

For the Fer expansion, it can be seen from Eqs. (37) - (42) that, the calculation of the obtained results in Eqs. (47) - (54), involves a series of nested commutators resulting in Eq. (42). The results of obtained for chemical shift, dipolar, and quadrupolar interactions might lead to the average Hamiltonian, $\overline{H}$, in the sense of the Magnus expansion under the following circumstances [43]:

$$\frac{F_1(\tau_C)}{\tau_C} = \overline{H} \tag{43}$$

where is the cyclic time (period of H). The difference between the Magnus and Fer expansions lies in the form of the correction terms. The result of Eq. (38) and the knowledge of the Hamiltonian H(t) in the Fer approach leads to a straightforward calculation of Eq. (42). $H_F^{(1)}$ is the time integral function which is used in the calculation of F_2. The iteration process can continue easily when the initial values of and are found. One major advantage of the Fer expansion over the AHT (Magnus expansion) is that only an evaluation of nested commutators is required in the calculation of (Eq. (42)). The Magnus expansion requires the calculation of nested commutators and their integrals to obtain the correction terms of a Hamiltonian. Blanes et al. proved the convergence of the Fer expansion and showed that the convergence of the Fer expansion is much faster than that of Magnus expansion [42, 50, 51, 54]. Madhu and Kurar also highlighted the observations that the calculation of a term like $H_F^{(1)}$ will contain several of the important signatures of the various higher-order terms in Magnus expansion, where all terms need to be calculated independently [39]. In addition, they mentioned that, the calculation of the infinite number of commutators in Eq. (42), although looking imposing, may turn out to be simpler to handle in most experimentally interesting cases due to the fast convergence and the negligible value of many of the commutators. Both approaches (Fer and AHT) may be complimentary and provide solutions to the time-dependent Schrodinger equation:

$$\frac{dU}{dt} = -iH(t)U(t) \tag{44}$$

However, the Fer expansion expresses the propagator in the form of an infinite-product of a series of exponentials given by:

$$U(t) = \prod_{k=1}^{\infty} \exp\left\{H_{F_k}(t)\right\} = \exp\left\{H_{F_1}(t)\right\}\exp\left\{H_{F_2}(t)\right\}.. \qquad (45)$$

IV.2. Applications

IV.2.1. Block-Siegert Shift

In the following the arguments given by Haeberlen [125,126], the untruncated rotating frame Hamiltonian during an on-resonance RF irradiation with is giving by

$$H = H_R^{on}(t) = -\Delta\omega I_Z - \omega_1\left[I_X \cos\phi + I_Y \sin\phi\right] - \omega_1\left[I_X \cos(2\omega_I t - \phi) + I_Y \sin(2\omega_I t - \phi)\right]$$

(46)

where ω_I is the Larmor frequency of the Ispins. $\Delta\omega$ is the off-resonance value, and is the RF nutation frequency. The period of $H_R^{on}(t)$ is $\tau_C = \dfrac{\pi}{\omega_I}$. A direct integration of Eq. (38) gives

$$F_1(t) = \int_0^t H dt = -\Delta\omega I_Z t - \omega_1 t\left[I_X \cos\phi - I_Y \sin\phi\right]$$

$$-\frac{\omega_1}{2\omega_I}\left\{I_X\left[\sin(2\omega_I t - \phi) + \sin\phi\right] + I_Y\left[\cos(2\omega_I t - \phi) - \cos\phi\right]\right\}$$

(47)

From Eq. (42), we obtain the following for with

$$H_F^{(0)} = H \qquad (48)$$

$$H_F^{(1)} = \sum_{k=1}^{\infty} \frac{(-1)^k k}{(k+1)!}\left\{F_1^k(t), H\right\} \qquad (49)$$

The first commutator of Eq. (42) is $-\frac{1}{2}\left[H, F_1(\tau_C)\right]$. The calculated coefficient term of ($-I_Z$) is found to be the Bloch-Siegert shift, given by

$$\omega_I(\frac{\omega_I}{2\omega_I})^2 \tag{50}$$

IV.2.2. Heteronuclear Dipolar Decoupling

Following the work of Haeberlen [125] and Ernst [126], we write the Hamiltonian in the doubly rotating frame of the I spin as

$$H(t) = d_{IS}S_Z(I_Z \cos\omega_{1I}t - I_Y \sin\omega_{1I}t) - \Delta\omega(I_Z \sin\omega_{1I}t - I_Y \cos\omega_{1I}t) \tag{51}$$

where d_{IS} is the dipolar coupling constant. $\Delta\omega$ is the off-resonance value, and is the nutation frequency of the RF irradiation on the I spins. $H(t)$ is periodic with period $\tau_C = \frac{2\pi}{\omega_{1I}}$. A direct integration of Eq. (38) gives

$$\begin{aligned} F_1(t) = \int_0^t Hdt = \frac{1}{\omega_{1I}} d_{IS}S_Z(I_Z \sin\omega_{1I}t + I_Y \cos\omega_{1I}t) \\ - \frac{\Delta\omega}{\omega_{1I}}(I_Z \cos\omega_{1I}t + I_Y \sin\omega_{1I}t) + \frac{1}{\omega_{1I}} d_{IS}S_Z I_Y - \frac{\Delta\omega}{\omega_{1I}} I_Y \end{aligned} \tag{52}$$

Using Eq. (52) for and evaluating the first commutator given by $-\frac{1}{2}[H, F_1(\tau_C)]$, the following result was obtained by both Madhu et al. [39] and Haeberlen [125]:

$$H_F^{(1)} \cong -\frac{1}{2\omega_{1I}} d_{IS}^2 S_Z^2 I_X + \frac{\Delta\omega}{\omega_{1I}} d_{IS}S_Z I_X - \frac{\Delta\omega^2}{2\omega_{1I}} I_X \tag{53}$$

Under on-resonance conditions, Madhu et al. [39] obtained the following expression of for the second commutator in Eq. (42)

$$H_F^{(1)} \cong -\frac{1}{2\omega_{1I}} d_{IS}^2 S_Z^2 I_X + \frac{1}{6}\frac{d_{IS}^3 S_Z^3}{\omega_{1I}^2} I_Z \tag{54}$$

The second term on the right-hand side in the above equation is obtained within the first-order correction in the case of Fer expansion, while in the case of AHT, this term corresponds to the second-order correction term.

IV.3. Advantages and Limitations

The Fer expansion [127-132] is an elegant expansion scheme involving the product of exponentials for the calculation of effective propagators and effective Hamiltonians in time-dependent problems which are of common occurrences in solid-state NMR.

Unlike in the Magnus expansion where an evaluation of nested commutators and their integrals are required to obtain the correction terms of a Hamiltonian, in the Fer expansion only an evaluation of nested commutators is required. Calculation of the infinite number of commutators, although looking imposing, may turn out to be simpler to handle in most experimentally interesting cases due to the fast convergence and the negligible value of many of the commutators. The Fer expansion can also be used as a numerical method for solving the time dependent Schrodinger equation.

V. Floquet-Magnus Expansion

Very recently, Mananga and Charpentier introduced the Floquet Magnus expansion approach to solid state NMR and spin physics [41]. This approach is unique in spin physics and useful in shedding new lights on AHT and FLT [41, 42,133,134]. The Floquet-Magnus expansion is a new theoretical tool for describing spin dynamics in solid-state NMR. This approach (FME) is an extension of the popular Magnus Expansion and average Hamiltonian theory. Calculations could be performed in a finite-dimensional Hilbert space instead of an infinite dimensional space within the Floquet theory. The FME approach is useful for controlling spin dynamic systems in solid-state NMR and evolves in the desired Lie group. All three theoretical approaches (AHT, FLT, and FME) are equivalent in the first order which corresponds to the popular average Hamiltonian.

$$H_{AHT}^{(0)} = H_{eff(FT)}^{1} = H_{1(FME)} = H_0 . \quad (55)$$

The FME approach can be considered as an improved AHT or a new version of FLT that could be very useful in simplifying calculations and in providing a more intuitive understanding of spin dynamics processes. The approach of FME is distinguished from other theories with its unique function ($n = 1,2,3,..$). This function provides an easy and alternative way for evaluating the spin behavior in between the stroboscopic observation points. The relationship to the regular Magnus expansion can be obtained from [41]

$$\frac{\Omega(T)}{T} = e^{-i\Lambda(0)} F e^{i\Lambda(0)} . \quad (56)$$

Eq. (56) shows that it is only in the case

$$\Lambda(0) = 0 \quad (57)$$

that the FME gives the AHT as provided by the ME. Eq. (56) is the general approach of the AHT called FME which gives also the option of $\Lambda(0) \neq 0$. The function $\Lambda_n(t)$ is connected to the appearance of features like spinning sidebands in MAS. The FME general formula is given by [41]

$$\Lambda_n(t) = \Lambda_n(0) + \int_0^t G_n(\tau) d\tau - tF_n \quad (58)$$

with

$$F_n = \frac{1}{T} \int_0^T G_n(\tau) d\tau \quad (59)$$

where and

$$G_1(\tau) = H(\tau) \quad (60)$$

$$G_2(\tau) = -\frac{i}{2}\left[H(\tau) + F_1, \Lambda_1(\tau)\right] \tag{61}$$

$$G_3(\tau) = -\frac{i}{2}\left[H(\tau) + F_1, \Lambda_2(\tau)\right] - \frac{i}{2}\left[F_2, \Lambda_1(\tau)\right] - \frac{1}{12}\left[\Lambda_1(\tau), \left[\Lambda_1(\tau), H(\tau) - F_1\right]\right] \tag{62}$$

The $\Lambda_n(t)$ functions ($n = 1,2,3,...$) represent the nth order term of the argument of the operator that introduces the frame such that the spin system operator is varying under the time independent Hamiltonian F. The evaluation of $\Lambda_n(t)$ is useful in many different ways, for example, in rotating NMR experiments of this function can be used to quantify the level of productivity of double quantum terms.

When using the Floquet-Magnus expansion, we considered a sample-spinning experiment. Indeed, the concept of AHT was found to be less descriptive for rotating systems, and these types of experiments were found to be more conveniently described using the Floquet theory [135]. Like the FLT, the FME describes the time evolution of the spin system at all times. The FME is more useful in simplifying calculations and in providing a greater understanding of spin dynamics processes. We showed that the lowest-order term as provided by AHT, FLT, and FME are all identical [41]. Next, using the FME, we calculated for each interaction, the first order function ($\Lambda_1(t)$) that provide an easy way for evaluating the spin system evolution. The function $\Lambda_1(t)$ is connected to the appearance of features like spinning sidebands in MAS. The evaluation of $\Lambda_1(t)$ is useful especially for the analysis of the non-stroboscopic evolution. We also showed that the second order ($\Lambda_2(t)$) is small in comparison to the first order ($\Lambda_1(t)$), and will be less useful in many cases. The function $\Lambda_1(t)$, available only in the FME scheme, will be useful in describing the spin dynamics in solid-state NMR and in understanding different synchronized or non-synchronized experiments.

V.1. Useful Equations and Propagators

Considering the initial conditions to be (i.e., $P(0) = I$), Blanes and co-workers derived explicitly, the first set of Floquet-Magnus expansion equations [42, 45, 50]. These equations are now considered to be a particular case ($\Lambda(0) = 0$) of a more general representation of the FME with [41]. The general formula for the contribution of the FME is given as by:

$$\Lambda_n(t) = \Lambda_n(0) + \int_0^t G_n(x)dx - tF_n \tag{63}$$

with

$$F_n = \frac{1}{T}\int_0^T G_n(x)dx \tag{64}$$

The first order contributions to the FME give

$$G_1(x) = H(x) \tag{65}$$

$$G_2(x) = -\frac{i}{2}\left[H(x) + F_1, \Lambda_1(x)\right] \tag{66}$$

$$G_3(x) = -\frac{i}{2}\left[H(x) + F_1, \Lambda_2(x)\right] - \frac{i}{2}\left[F_2, \Lambda_1(x)\right] - \frac{1}{12}\left[\Lambda_1(x), \left[\Lambda_1(x), H(x) - F_1\right]\right] \tag{67}$$

Calculation software can enable formal derivation of higher order terms. In Eqs. (63-64) above, the functions with, represent the nth order term of the argument of the operator that introduces the frame such that the spin system operator is varying under the time independent Hamiltonian F .

The function ($\Lambda_n(t)$) can be useful to quantify the level of productivity of double quantum terms [45, 46]. The FME propagator is given by:

$$U(t) = P(t)\exp\{-itH_F\}P^+(0) \tag{68}$$

Here the constraint of stroboscopic observation is removed. $P(t)$ is the operator that introduces the frame that varies under the time independent Hamiltonian H_F. The function given above is the argument of the operator $P(t)$ such that:

$$P(t) = \exp\{-i\Lambda(t)\} \tag{69}$$

V.2. Applications

V.2.1. Simple Cases

Common Form of Hamiltonian in Solid-State NMR

We revisited the static perturbation theory [41, 61, 136] which has been shown to yield the correct form of Zeeman truncated NMR interactions without the limit of stroboscopic observation of the AHT. This gave us the opportunity to shed light on the FME scheme and the derivation of a criterion for the two theories being compatible.

For the sake of simplicity (but without loss of generality), we considered the Hamiltonian

$$H = \omega_0 I_Z + \lambda \sum_m (-1)^m R_{2,-m} T_{2,+m} \tag{70}$$

This is a common form of Hamiltonian in solid-state NMR is the Zeeman interaction, $R_{2,m}$ are the lattice parts of the internal interaction which encode its orientational dependence with respect to the magnetic field, $T_{2,m}$ are second rank m-order spherical tensors describing the spin system as defined by $\left[I_Z, T_{2,m}\right] = mT_{2,m}$. The static perturbation theory (SPT) in terms of the irreducible tensor operators gives the diagonal Hamiltonian (with respect to $\omega_0 I_Z$)

$$H_{SPT} = \omega_0 I_Z + \lambda R_{2,0} T_{2,0} + \frac{\lambda^2}{2\omega_0} \sum_{m \neq 0} \frac{R_{2,m} R_{2,-m}}{m} \left[T_{2,m}, T_{2,-m}\right] \tag{71}$$

As discussed in the seminal work [61], discrepancies between AHT and FT appear in the rotating frame representation (or more generally in the interaction frame) where the Hamiltonian becomes time-dependent:

$$\widetilde{H}(t) = e^{+i\omega_0 I_z t} H e^{-i\omega_0 I_z t} = \lambda \sum_m (-1)^m R_{2,-m} T_{2,m} e^{im\omega_0 t} \tag{72}$$

The FME expansion Eqs. (63) and (64) yield as first order terms:

$$F_1 = \lambda R_{2,0} T_{2,0} \tag{73}$$

$$\Lambda_1(t) = \lambda \sum_{m\neq 0} (-1)^m \frac{R_{2,-m} T_{2,+m}}{im\omega_0} e^{im\omega_0 t} \tag{74}$$

whereas the AHT (stroboscopic detection) yields [41]

$$\Lambda_1(t) = \lambda \sum_{m\neq 0} (-1)^m \frac{R_{2,-m} T_{2,+m}}{im\omega_0} (e^{im\omega_0 t} - 1) \tag{75}$$

Also, the FME scheme provides the same second order term as in SPT theory

$$F_2 = \frac{\lambda^2}{2\omega_0} \sum_{m\neq 0} \frac{R_{2,m} R_{2,-m}}{m} \left[T_{2,m}, T_{2,-m} \right] \tag{76}$$

This shows that FME provides an expansion in the rotating frame which is in agreement with the static perturbation theory and the Van Vleck transformations. This is not the case of the Magnus expansion. This agreement can be explained by the connection that exists between the FME and secular pertubation theory (SPT) propagators as explained in the original article [41].

Extension to Multimode Hamiltonian

The application of FME to multimode Hamiltonian with frequencies $\vec{\omega} = (\omega_1, \ldots, \omega_N)$ is straightforward. Considering the generalized Fourier expansion of the Hamiltonian (where the frequency indices are represented by):

$$H(t) = \sum_{\vec{m}} H_{\vec{m}} \exp(-i\vec{m}.\vec{\omega}t) \tag{77}$$

we obtain

$$\Lambda_1(t) = \sum_{\vec{m}.\vec{\omega}\neq 0} \frac{H_{\vec{m}}}{i\vec{m}.\vec{\omega}} e^{-i\vec{m}.\vec{\omega}t} \tag{78}$$

and

$$F_1 = \sum_{\vec{m}.\vec{\omega}=0} H_{\vec{m}} \tag{79}$$

Similarly, calculation of second order terms is straightforward [41]. These expressions highlight the fact that the multimode Hamiltonian case can be easily treated in Hilbert space with the FME.

V.2.2. Investigation of the Effect of Finite Pulse Errors on the BABA Pulse Sequence

In Figure 2, the relation is valid only during the interval where $\theta(t)$ acted.

We studied the finite pulse widths effects for the BABA pulse sequence using the Floquet-Magnus expansion approach [46,137]. We investigated the case $m = 1$ and considered a system of 2 spins. Only DQ terms are considered for the function $\Lambda_1(t)$. We considered the simple case where the rotations are: $\alpha^{ij} = \beta^{ij} = \gamma^{ij} = 0$.

The coefficient C_n^{ij} are $C_1 = \frac{-1}{\sqrt{3}}\sin(\theta)\cos(\theta)e^{-i\phi}$, $C_{-1} = 0$, $C_2 = \frac{1}{(2\sqrt{6})(\sin^2(\theta)e^{-2i\phi})}$, $C_{-2} = 0$. For example, with $\theta = \frac{\pi}{4}$ and, $\phi = 0$ the coefficients are $C_1 = \frac{-1}{2\sqrt{3}}$ and $C_2 = \frac{1}{4\sqrt{6}}$. The function $\Lambda_1(t)$ is written as:

$$\Lambda_1(t) = \frac{3}{2\sqrt{6}} b_{ij} \left[a_{-1} C_1 (\frac{\tau + 4\tau_p}{2})\psi + a_{-2} C_2 (\frac{\tau_R + 4\tau_p}{2})\psi \right] (I_{YY} - I_{XX}) \tag{80}$$

where

$$\psi = \frac{t}{\tau_R} \tag{81-a}$$

$$\phi = \frac{2\tau_p}{\tau_R} \tag{81-b}$$

$$a_{-1} = \frac{-1}{2\pi i} e^{-i\pi(1+\phi)} \left[e^{-i\pi(1-2\phi)} - 1 \right] \tag{81-c}$$

and

$$a_{-2} = \frac{-1}{4\pi i} e^{-i2\pi(1+\phi)} \left[e^{-i2\pi(1-2\phi)} - 1 \right] \tag{82}$$

which lead to

$$\frac{\Lambda_1(\psi, \phi)}{b_{ij}\tau_R} = \frac{3}{2\sqrt{6}} (-a_{-1} + \frac{1}{2\sqrt{2}} a_{-2})(\frac{1}{2} + \phi)\psi (I_{YY} - I_{XX}) \tag{83}$$

As reported in the literature [46], we considered the case, $0.1 \le \phi \le 0.606$, which corresponds to the spinning frequencies $\frac{\omega_R}{2\pi} = 5 - 10$ kHz and to the recoupling RF fields kHz. We generated two types of plots from Eq. (83). First, the plot of $\frac{\Lambda_1(t)}{b_{ij}\tau_R}$ versus $\psi = \frac{t}{\tau_R}$, while keeping $\phi = \dfrac{2\tau_p}{\tau_R}$ constant corresponds to Figure 3. The plot of $\dfrac{\Lambda_1(t)}{b_{ij}\tau_R}$ versus $\phi = \dfrac{2\tau_p}{\tau_R}$, while keeping the time t constant corresponds to Figure 4.

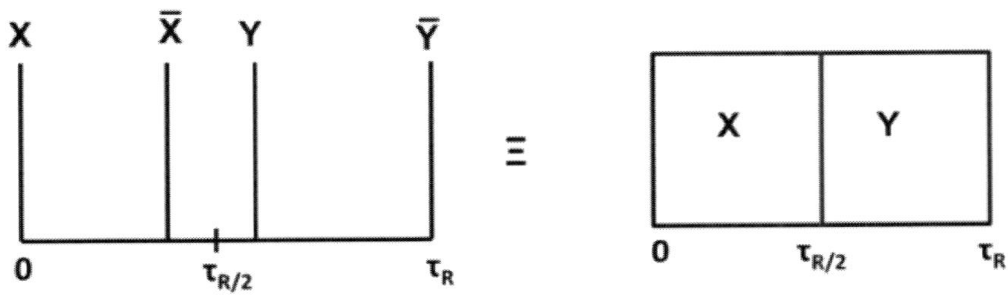

Figure 1. BABA pulse sequence with $\delta - pulse$ width.

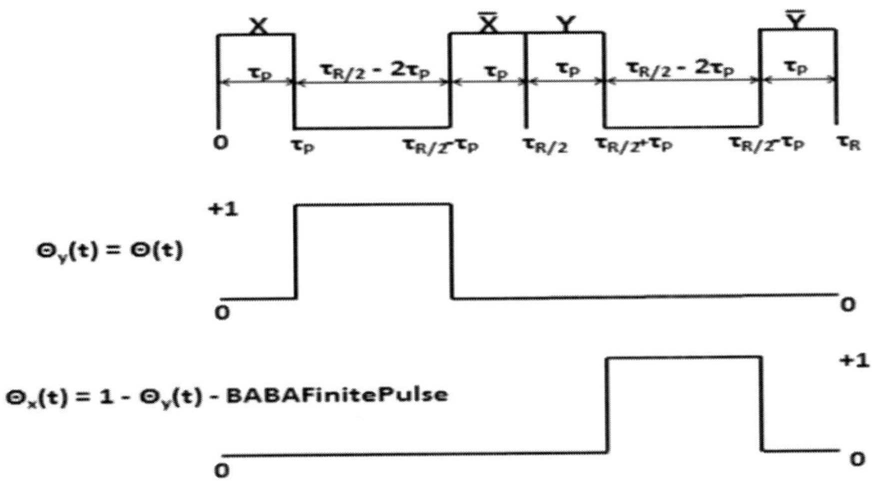

Figure 2. BABA pulse sequence with finite pulse width.

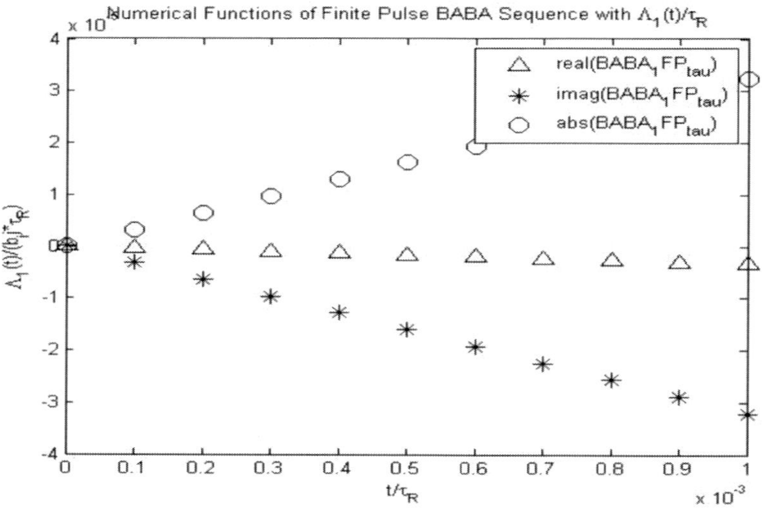

a

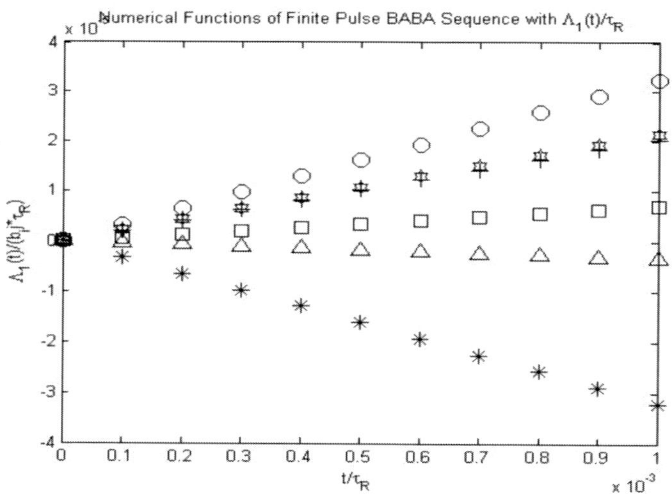

b

a) For $\phi = 0.1$ b) For $\phi = 0.1$ and $\phi = 0.606$

Figure 3. Numerical Functions of Finite Pulse BABA Sequence with $\dfrac{\Lambda_1(t)}{b_{ij}\tau_R}$ versus $\psi = \dfrac{t}{\tau_R}$.

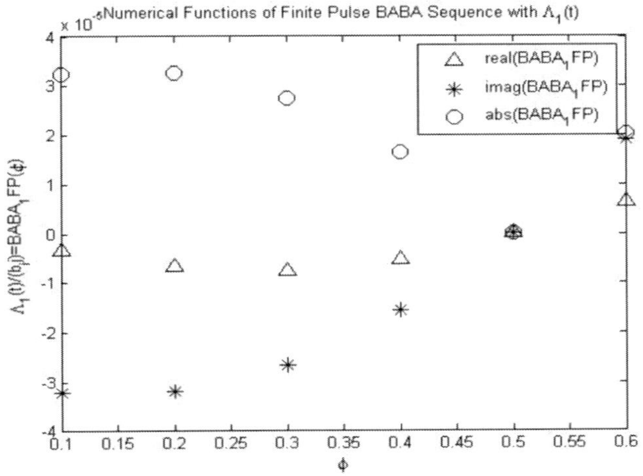

a

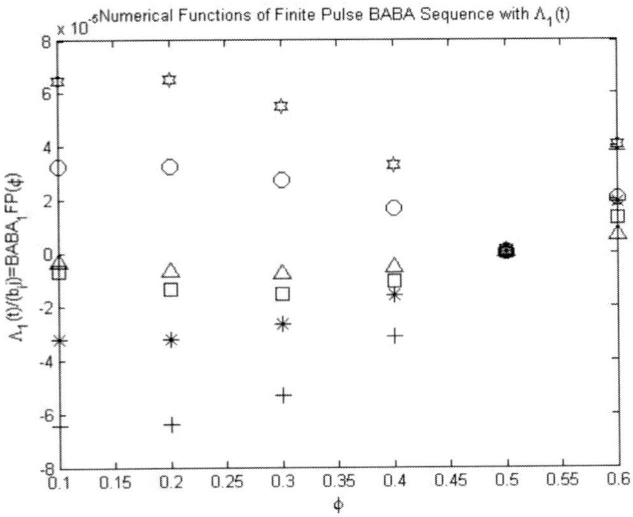

b

a) For t = 1 ms b) For t = 1 ms and t = 2 ms

Figure 4. Numerical Functions of Finite Pulse BABA Sequence with $\frac{\Lambda_1(t)}{b_{ij}}$ versus

$\phi = \frac{2\tau_p}{\tau_R}$.

Analysis of Figures 3 and 4

Figure 3 a) shows the dimensionless function for BABA pulse sequence with finite pulse widths versus the dimensionless number $\psi = \frac{t}{\tau_R}$, for $\phi = 0.1$. Figure 3 b) shows the plot of the same function versus $\psi = \frac{t}{\tau_R}$, for the two cases: and $\phi = 0.606$. Due to the complexity of the function $\frac{\Lambda_1(t)}{b_{ij}\tau_R}$, real, imaginary and absolute parts are plotted separately as a function of ψ. In Figure 3 b), the symbols "square", "plus" and "hexagram" represent respectively the real, imaginary, and absolute parts of the function for $\phi = 0.606$. These functions depend on the DQ terms. Therefore, the study of the amplitude of DQ terms can be considered as a viable approach for controlling the complex spin dynamics of a spin system evolving under the dipolar interaction of the BABA pulse sequence with finite widths. The plot can be considered as a quantitative representation of the amplitude of the DQ coherence as a function of ψ. The size of determine the amplitude of the DQ coherence, which indicates the degree of efficiency of the scheme. A closer look at Figs. 3 and 4 (BABA with finite pulse widths) comparing BABA with delta-pulse sequences [45] shows that the magnitude of the DQ terms of BABA with finite pulses is, as expected, small compared to the magnitude of BABA with δ pulse sequences as shown in Eq. (84)

$$\left|\frac{\Lambda_1(t)}{b_{ij}\tau_R}\right|_{finite-pulse} < \left|\frac{\Lambda_1(t)}{b_{ij}\tau_R}\right|_{\delta-pulse} \tag{84}$$

Figure 4 a) shows the plot of the function for the BABA pulse sequence with finite pulse widths versus the dimensionless number $\phi = \frac{2\tau_p}{\tau_R}$, for $t = 1$ ms.

Figure 4 b) shows the plot of the same function versus for the two cases: ms and $t = 2$ ms. As with Figure 3, due to the complexity of the function

$\frac{\Lambda_1(t)}{b_{ij}}$, real, imaginary, and absolute parts are plotted separately as function of ϕ.

In Figure 4 b), the symbols “square”, “plus”, and “hexagram” represent respectively the real, imaginary, and absolute parts of the function for ms. These functions depend on the DQ terms. It can easily be seen that, as expected when $\phi = \frac{2\tau_p}{\tau_R}$ increases, the magnitude of the double quantum terms decreases. When $\phi \to 0$, the magnitude of the DQ term → maximum corresponding to the delta-pulse sequence. However, when $\phi = 0.5$ corresponding to $\tau_p = \frac{\tau_R}{4}$, we have $\frac{\Lambda_1(t)}{b_{ij}} = 0$.

The strength of the DQ terms decreases, cancels and builds up again. This dynamic predicts that a full decoupling is possible, which occurs at $\phi = 0.5$. The plot of the magnitude of the double quantum term of $\Lambda_1(t)$ as a function of the pulse length gives a basic understanding of the experiment, such as how to select robust finite pulse widths and how to select finite pulse widths that maximize or minimize double quantum terms. The study of this function could be helpful in predicting the conditions of decoupling.

V.2.3. Study of Interactions in Solid-State NMR When Irradiated with the Magic Echo Pulse Sequence

Let us first describe various interactions solid-state NMR. The interactions of spin systems in NMR have been extensively described by Levitt [18], Veshtort and Griffin [19], and Bak et al. [20]. In this part of the chapter, we summarize the various forms of Hamiltonians subject to the secular approximation of high-field truncation. The Hamiltonian of the spin system is the sum of the different physical influence or interaction acting on the nuclear spin such as the chemical shift, dipole-dipole coupling, or quadrupolar coupling. Mathematically, the spin Hamiltonian in NMR can be written as

$$\mathrm{H} = \sum_{\Lambda} \mathrm{H}^{\Lambda} \tag{85}$$

where represents the contributions from radiofrequency (RF) irradiation, chemical shift (CS), indirect spin-spin coupling (J), dipole-dipole coupling (D), quadrupolar coupling (Q) and so on. In general, the interactions of the nuclei with the excitation field (external radiation) and the internal Hamiltonians (CS, J, D, Q) do not commute.

In high magnetic field NMR, all interaction terms have the spin quantum numbers equal to zero ($\mu = 0$, $m = 0$) and the spin Hamiltonians are truncated by the secular approximation. The NMR interactions are written by the following expressions

$$H_{RF} = \sum_i \left|\omega_{RF}^i(t)\right|(I_{ix}\cos\phi_i + I_{iy}\sin\phi_i) \tag{86}$$

$$H_{CS} = \sum_i \omega_{CS,0}^i(t) I_{iz} \tag{87}$$

$$H_J = \sum_{ij} -\omega_{J_{iso,0}}^{ij}(t)\frac{1}{\sqrt{3}}\mathrm{I}_i.\mathrm{I}_j + \omega_{J_{aniso,0}}^{ij}(t)\frac{1}{\sqrt{6}}(3I_{iz}I_{jz} - \mathrm{I}_i.\mathrm{I}_j) \tag{88}$$

$$H_D = \sum_{i,j} \omega_{D,0}^{ij}(t)\frac{1}{\sqrt{6}}(3I_{iz}I_{jz} - \mathrm{I}_i.\mathrm{I}_j) \tag{89}$$

$$H_Q = \sum_i \omega_{Q,0}^i(t)\frac{1}{\sqrt{6}}(3I_{iz}^2 - \mathrm{I}_i^2) + \frac{1}{2\omega_0^i}\{\omega_{Q,-2}^i(t)\omega_{Q,2}^i(t)(2\mathrm{I}_i^2 - 2I_{iz}^2 - 1)I_{iz})$$
$$+ \omega_{Q,-1}^i(t)\omega_{Q,1}^i(t)(4\mathrm{I}_i^2 - 8I_{iz}^2 - 1)I_{iz}\} \tag{90}$$

with representing the concerned spins. is the phase of the RF irradiation (pulse phases), and is the RF nutation frequency. (CS) represents the chemical shift, (J) is the indirect spin-spin coupling, (D) represents the dipole-dipole coupling, and (Q) is the quadrupolar coupling. The symbols is the gyromagnetic ratio, and describe scalar and anisotropic J coupling, respectively. Details on the NMR interactions, shown in Eqs. (86) - (90), can be found in the literature [4, 5, 18-20, 34,139].

The operator product is truncated to when a coupling occurs between nuclei of different spin species. In the Eq. (90), the first term represents the first-order quadrupolar coupling while the second terms include the secular components for the second-order quadrupolar coupling. In the following, we have neglected the second-order term with the secular components. Eq. (90) is reduced to

$$H_Q = \sum_i \omega_{Q,0}^i(t)\frac{1}{\sqrt{6}}(3I_{iz}^2 - \mathrm{I}_i^2) \tag{91}$$

The frequency coefficients for the various internal Hamiltonians depend on some physical parameters and are typically expressed in terms of Fourier expansion such as

$$\omega_{\lambda,m'}(t) = \sum_{m=-2}^{2} \omega_{\lambda,m'}^{(m)} e^{im\omega_r t} \tag{92}$$

where is the spin rate. The Fourier coefficients are written as

$$\omega_{\lambda,m'}^{(m)} = \omega_{iso}^{\lambda}\delta_{m,0} + \omega_{aniso}^{\lambda}\left\{D_{0,-m}^{(2)}(\Omega_{PR}^{\lambda}) - \frac{\eta^{\lambda}}{\sqrt{6}}\left[D_{-2,-m}^{(2)}(\Omega_{PR}^{\lambda}) + D_{2,-m}^{(2)}(\Omega_{PR}^{\lambda})\right]\right\}d_{-m,m'}^{(2)}(\beta_{RL}) \tag{93}$$

where is the standard Kronecker delta. The constants describing the isotropic (ω_{iso}^{λ}) and the anisotropic ($\omega_{aniso}^{\lambda}, \eta^{\lambda}$) contributions to the Fourier coefficients can be found in the article by Bak et al. [20]. That article also gives a description of the orientation dependence for the anisotropic interactions. The frames P, R, C, and L represent the principal-axis, the rotor-fixed, the crystal-fixed and the laboratory-fixed frames respectively. These frames (P, R, S, and L) are connected by the relation

$$D_{m',m}^{(2)}(\Omega_{PR}^{\lambda}) = \sum_{m''=-2}^{2} D_{m',m''}^{(2)}(\Omega_{PC}^{\lambda})D_{m'',m}^{(2)}(\Omega_{CR}) \tag{94}$$

The internal Hamiltonians (Eqs. (86)-(90)) can also be written in more succinct forms such as those described by Tycko [121] or Schnell and Spiess [4]. For instance, in the form given by Tycko, the Hamiltonian is expressed in terms of $\{\mathrm{Y}_{2m}\}$, which is a basis for an irreducible representation of rotations of the spatial coordinates of the sample, and $\{T_{2m}\}$, which is a basis for an irreducible representation of rotations of the spin angular momenta. and are second-rank spherical harmonics, and second-rank irreducible tensor operators, respectively. In high field approximation, we only have the two basis and $\{T_{20}\}$. For example, the truncated dipole-dipole coupling can be written as

$$H_D = b_{ij} Y_{20} T_{20} \tag{95}$$

where is the coupling constant proportional to r_{ij}^{-3} (and is the internuclear distance). The quadrupolar interaction can also be written as

$$H_Q^{(0)}(t) = \sum_{m=-2}^{2} \left\{ \frac{1}{\sqrt{6}} D_{m0}^{(2)}(\Omega) + \eta \left[D_{m2}^{(2)}(\Omega) + D_{m-2}^{(2)}(\Omega) \right] \right\} D_{0m}^{(2)} \left[\Omega'(t) \right] T_{20} \tag{96}$$

where η is the asymmetry parameter.

In the following analysis, we used the FME under the combined presence of MAS and RF irradiation, with their cycle times being non-commensurate. Considering the representation that govern the evolution time in the magic echo pulse sequence given in reference [44] we obtain the following results:

For Chemical Shift (CS)

$$F_{1(CS)} = \overline{H}_{\Delta\omega}^{0} = \frac{1}{3\tau} \frac{\Delta\omega}{\omega_{RF}} I_y [1 - \cos\omega_{RF}(2\tau - \alpha)] - \frac{1}{3\tau} \frac{\Delta\omega}{\omega_{RF}} I_x [\sin\omega_{RF}(2\tau - \alpha) + \frac{4}{\pi}\omega_{RF}\alpha] \tag{97}$$

$$\Lambda_{1(CS)}(t) = \sum_{n=-2}^{+2} C_n \Big[-(I_A + 2I_B + I_C) I_x + (I_B + I_D) I_y + \Big(\big(\frac{1}{in\omega_R}\big)(e^{-in\omega_R t} - 1)(1 - 2a_0) + \sum_{m=-\infty}^{\infty} \big(\frac{-2}{im\omega_R}\big) a_{m-n} (e^{-im\omega_R t} - 1) \Big) I_z \Big] - tF_{1(CS)} \tag{98}$$

where I_A, I_B, I_C, and are given by:

$$I_A = \int_0^t e^{-in\omega_R t'} \cos(\omega_{RF} t')dt' = \frac{1}{(n^2\omega_R^2 - \omega_{RF}^2)} \left[-in\omega_R + in\omega_R \exp(-in\omega_R t)\cos(\omega_{RF} t) + -\omega_{RF} \exp(-in\omega_R t)\sin(\omega_{RF} t) \right] \quad (99)$$

$$I_B = \int_0^t e^{-in\omega_R t'} \sin(\omega_{RF} t')dt' = \frac{1}{(n^2\omega_R^2 - \omega_{RF}^2)} \left[-\omega_{RF} + \omega_{RF} \exp(-in\omega_R t)\cos(\omega_{RF} t) + +in\omega_R \exp(-in\omega_R t)\sin(\omega_{RF} t) \right] \quad (100)$$

$$I_C = \int_0^t e^{-in\omega_R t'} \cos\omega_{RF}(t' - \xi)dt' = -\frac{1}{(n^2\omega_R^2 - \omega_{RF}^2)} \left[in\omega_R \cos(\omega_{RF}\xi) + \omega_{RF} \sin(\omega_{RF}\xi) + -in\omega_R \exp(-in\omega_R t)\cos(-\omega_{RF} t + \omega_{RF}\xi) - \omega_{RF} \exp(-in\omega_R t)\sin\omega_{RF}(\xi - t) \right] \quad (101)$$

$$I_D = \int_0^t e^{-in\omega_R t'} \sin\omega_{RF}(t' - \xi)dt' = \frac{1}{(n^2\omega_R^2 - \omega_{RF}^2)} \left[in\omega_R \sin(\omega_{RF}\xi) - \omega_{RF} \cos(\omega_{RF}\xi) + -in\omega_R \exp(-in\omega_R t)\sin\omega_{RF}(\xi - t) + \omega_{RF} \exp(-in\omega_R t)\cos\omega_{RF}(\xi - t) \right] \quad (102)$$

with

$$2\omega_{RF}\alpha = \frac{\pi}{2} \quad (103)$$

and

$$\xi = 2\tau - \alpha \quad (104)$$

For Dipolar Coupling (D)

$$F_{1(D)} = \overline{\mathrm{H}}_D^0 = \omega_D I_z I_z - \omega_D \mathrm{I.I} + \frac{1}{4\tau}\frac{\omega_D}{\omega_{RF}} I_x I_x [2\omega_{RF}(2\tau + \alpha)] + \sin 2\omega_{RF}(2\tau - \alpha)] + + \frac{1}{4\tau}\frac{\omega_D}{\omega_{RF}} I_y I_y [2\omega_{RF}(2\tau - \alpha)] - \sin 2\omega_{RF}(2\tau - \alpha)] - \frac{1}{2\tau}\frac{\omega_D}{\omega_{RF}}[I_x I_y + I_y I_x]\sin^2\omega_{RF}(2\tau - \alpha) \quad (105)$$

$$\Lambda_{1(D)}(t)=\sum_{n=-2}^{+2}C_n^{ij}\Big[3(I_E I_{iz}I_{jz}+I_F I_{ix}I_{jx})-\frac{3}{2}I_I(I_{ix}I_{jz}+I_{iz}I_{jx})+((\frac{1}{in\omega_R})(e^{-ix\omega_R t}-1)I.I+$$
$$3(I_E I_{ix}I_{jx}+I_F I_{iy}I_{jy})-\frac{3}{2}I_I(I_{ix}I_{jy}+I_{iy}I_{jx})+\frac{1}{in\omega_R}(e^{-in\omega_R t}-1)I.I+3(I_H I_{ix}I_{jx}+I_G I_{iy}I_{jy})$$
$$\frac{3}{2}I_J(I_{ix}I_{jy}+I_{jy}I_{jx})+\frac{1}{in\omega_R}(e^{-in\omega_R t}-1)I.I+3(I_E I_{iz}I_{jz}+I_F I_{ix}I_{jx})+\frac{3}{2}I_I(I_{ix}I_{jz}+I_{iz}I_{jx})+$$
$$+\frac{1}{in\omega_R}(e^{-in\omega_R t}-1)I.I-\frac{1}{in\omega_R}(e^{-im\omega_R t}-1)(3I_{iz}I_{jz}-I.I)\Big]-tF_{1(D)}$$
(106)

where I_E, I_F, I_G, I_H, and are given by:

$$I_E=\int_0^t e^{-in\omega_R t'}\cos^2(\omega_{RF}t')dt'=\frac{1}{2}\frac{\omega_R}{n(n^2\omega_R^2-4\omega_{RF}^2)}\Big[-2in^2\omega_R^2+4i\omega_{RF}^2+i\exp(-in\omega_R t)n^2\omega_R^2$$
$$+in^2\omega_R^2\exp(-in\omega_R t)\cos(2\omega_{RF}t)-2\omega_{RF}\exp(-in\omega_R t)\sin(2\omega_{RF}t)n\omega_R-4i\exp(-in\omega_R t)\omega_{RF}^2\Big]$$
(107)

$$I_F=\int_0^t e^{-in\omega_R t'}\sin^2(\omega_{RF}t')dt'=-\frac{1}{2}\frac{\omega_R}{n(n^2\omega_R^2-4\omega_{RF}^2)}\Big[-4i\omega_{RF}^2-i\exp(-in\omega_R t)n^2\omega_R^2$$
$$+in^2\omega_R^2\exp(-in\omega_R t)\cos(2\omega_{RF}t)-2\omega_{RF}\exp(-in\omega_R t)\sin(2\omega_{RF}t)n\omega_R+4i\exp(-in\omega_R t)\omega_{RF}^2\Big]$$
(108)

$$I_G=\int_0^t e^{-in\omega_R t'}\sin^2\omega_{RF}(t'-\xi)dt'=-\frac{1}{2}\frac{\omega_R}{n(n^2\omega_R^2-4\omega_{RF}^2)}\Big[2in^2\omega_R^2-4i\omega_{RF}^2-2in^2\omega_R^2\cos^2(\omega_R\xi)$$
$$-in^2\omega_R^2\exp(-in\omega_R t)-4\omega_{RF}n\omega_R\sin(\omega_{RF}\xi)\cos(\omega_{RF}\xi)+4i\exp(-in\omega_R t)\omega_{RF}^2+$$
$$n^2\omega_R^2\exp(-in\omega_R t)\cos 2\omega_{RF}(\xi-t)+$$
$$+2\omega_{RF}\exp(-in\omega_R t)\sin 2\omega_{RF}(\xi-t)n\omega_R+4i\exp(-in\omega_R t)\omega_{RF}^2\Big]$$
(109)

$$I_H=\int_0^t e^{-in\omega_R t'}\cos^2\omega_{RF}(t'-\xi)dt'=\frac{1}{2}\frac{\omega_R}{n(n^2\omega_R^2-4\omega_{RF}^2)}\Big[-2in^2\omega_R^2\cos^2(\omega_R\xi)+4i\omega_{RF}^2+$$
$$+in^2\omega_R^2\exp(-in\omega_R t)-4\omega_{RF}n\omega_R\sin(\omega_{RF}\xi)\cos(\omega_{RF}\xi)-4i\exp(-in\omega_R t)\omega_{RF}^2+$$
$$+in^2\omega_R^2\exp(-in\omega_R t)\cos 2\omega_{RF}(\xi-t)+2\omega_{RF}\exp(-in\omega_R t)\sin 2\omega_{RF}(\xi-t)n\omega_R\Big]$$
(110)

$$I_I = \int_0^t e^{-in\omega_R t'} \sin(2\omega_{RF}t')dt' = \frac{1}{(n^2\omega_R^2 - 4\omega_{RF}^2)}\left[-2\omega_{RF} + 2\omega_{RF}\exp(-in\omega_R t)\cos(2\omega_{RF}t) + \right.$$
$$\left. + in\omega_R \exp(-in\omega_R t)\sin(2\omega_{RF}t)\right] \tag{111}$$

$$I_J = \int_0^t e^{-in\omega_R t'} \sin 2\omega_{RF}(t'-\xi)dt' = \frac{-1}{(n^2\omega_R^2 - 4\omega_{RF}^2)}\left[-2\omega_{RF} + 4\omega_{RF}\cos^2(\omega_{RF}\xi) + \right.$$
$$-2in\omega_R \sin(\omega_{RF}\xi)\cos(\omega_{RF}\xi) - 2\omega_{RF}\exp(-in\omega_R t)\cos 2\omega_{RF}(\xi - t) +$$
$$\left. + in\omega_R \exp(-in\omega_R t)\sin 2\omega_{RF}(\xi - t)\right] \tag{112}$$

For Quadrupolar Interaction (Q)

Using the following magic echo pulse sequence scheme:

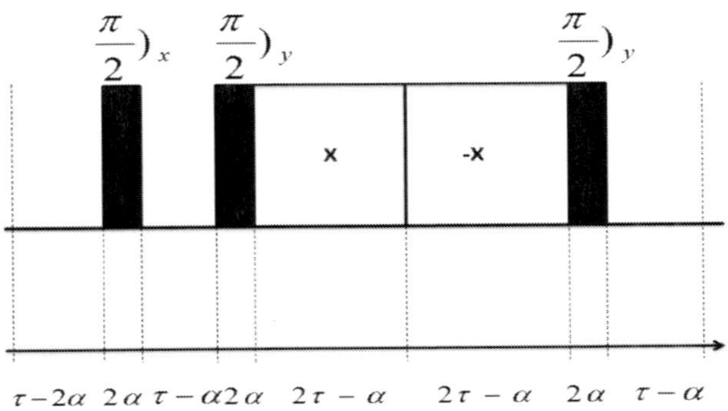

Figure 5. Magic echo sequence for refocusing the quadrupolar interaction.

we have:

$$F_{1(Q)} = \overline{\mathrm{H}}_{\omega_Q}^{0} = \frac{1}{7\tau}\frac{12\alpha\omega_Q}{\pi}I_{x,2} + \frac{4(3\alpha-\tau)\omega_Q}{7\tau}\left[I_{x,1}I_{x,1} + I_{y,1}I_{y,1} - 2I_{z,1}I_{z,1}\right] \tag{113}$$

The free evolution diagram in this case is given by figure 6:

In the above scheme (Figure 6), we have the following relation in each part:

$$a_k^{(1)} = \frac{1}{\tau_R} \int_0^{\tau-2\alpha} e^{ik\omega_R t'} dt' \tag{114}$$

$$a_0^{(1)} = \frac{1}{\tau_R}(\tau - 2\alpha) \tag{115}$$

$$a_k^{(1)} = \frac{1}{ik2\pi}\left[e^{ik\omega_R(\tau-2\alpha)} - 1\right] \tag{116}$$

$$a_k^{(2)} = \frac{1}{\tau_R}\left[\int_\tau^{2\tau-\alpha} e^{ik\omega_R t'} dt' + \int_{6\tau+\alpha}^{7\tau} e^{ik\omega_R t'} dt'\right] \tag{117}$$

$$a_0^{(2)} = \frac{1}{\tau_R}(2\tau - 2\alpha) \tag{118}$$

$$a_k^{(2)} = \frac{1}{ik2\pi}\left[e^{ik\omega_R(2\tau-\alpha)} - e^{ik\omega_R\tau} + e^{ik\omega_R 7\tau} - e^{ik\omega_R(6\tau+\alpha)}\right] \tag{119}$$

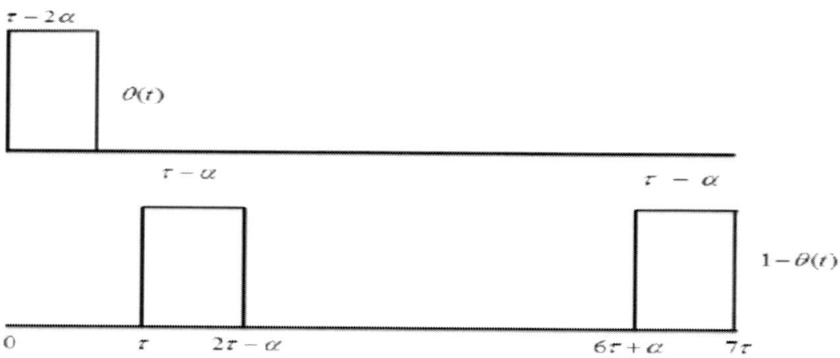

Figure 6. The relation $1-\theta(t)$ is only valid during the interval where acts.

However, for mathematical convenience, we made the following approximation: we neglected the first free evolution time and considered the free evolution time given in figure 7:

In this new dynamics picture, the toggling frame is written as

$$\widetilde{H}_{(Q)}(t) = \widetilde{H}_{(2)} + \widetilde{H}_{(3)}\theta(t) + \widetilde{H}_{(4)} + \widetilde{H}_{(5)} + \widetilde{H}_{(6)} + \widetilde{H}_{(7)} + \widetilde{H}_{(8)}(1-\theta(t)), \tag{120}$$

where $\widetilde{H}_{(2)}$, $\widetilde{H}_{(3)}$, $\widetilde{H}_{(4)}$, $\widetilde{H}_{(5)}$, $\widetilde{H}_{(6)}$, $\widetilde{H}_{(7)}$, and correspond to the toggling Hamiltonian for the intervals δ_2, δ_3, δ_4, δ_5, δ_6, δ_7, δ_8, respectively, given in the literature [43, 44]. Processing as previously, the first and second part of the sequence are:

$$a_k^{(1)} = \frac{1}{\tau_R}\int_{2\alpha}^{\tau+\alpha} e^{ik\omega_R t}\,dt \tag{121}$$

$$a_k^{(2)} = \frac{1}{\tau_R}\int_{5\tau+3\alpha}^{6\tau+2\alpha} e^{ik\omega_R t}\,dt \tag{122}$$

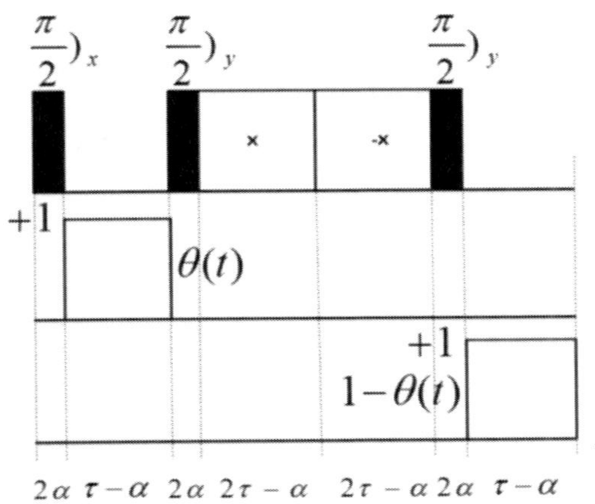

Figure 7. Magic echo sequence for refocusing the quadrupolar interaction with the first free evolution time shown in Figure 3, neglected.

From the above equations, we obtain the following criteria

$$a_0 = c - \phi \tag{123}$$

$$a_k = a_k^{(2)} = a_k^{(1)} e^{ik2\pi(5c+\phi)} = \frac{1}{2\pi ik}\left[e^{ik2\pi(c+\phi)} - e^{ik4\pi\phi}\right]e^{ik2\pi(5c+\phi)} \tag{124}$$

with

$$\phi = \frac{\alpha}{\tau_R} \tag{125}$$

and

$$\tau = c\tau_R \tag{126}$$

The first order of the FME is computed to be:

$$\begin{aligned}\Lambda_{1(Q)}(t) = \sum_{n=-2}^{+2} C_n^{(Q)} \big[& 12(I_E I_{z,1} I_{z,1} + I_F I_{y,1} I_{y,1}) + 6I_{z,2} I_I - 5(\frac{1}{in\omega_R})(e^{-in\omega_R t} - 1)I.I + 12I_E I_{y,1} I_{y,1} + \\ & + 12I_F I_{x,1} I_{x,1} - 6I_I I_{z,2} + 12I_{x,1} I_{x,1} I_E + 12I_F I_{z,1} I_{z,1} + 6I_I I_{y,2} + 12I_H I_{x,1} I_{x,1} + 12I_G I_{z,1} I_{z,1} \\ & - 6I_J I_{y,2} + 12I_E I_{x,1} I_{x,1} + 12I_F I_{y,1} I_{y,1} + 6I_{z,2} I_I + 12I_{y,1} I_{y,1}(-\frac{1}{in\omega_R})(e^{-in\omega_R t} - 1)\big] - tF_{1(Q)}\end{aligned} \tag{127}$$

Considering Figure 7, we remark that the magic echo pulse sequence remove the dependence and calculation of and become less lengthy.

V.3. Advantages and Limitations

The aims of the FME is to bridge the AHT to the Floquet Theorem but in a more concise and efficient formalism [41]. Calculations can then be performed in a finite-dimensional Hilbert space instead of an infinite dimensional space within the Floquet theory. We expect that the FME will provide means for devising new RF pulse sequence and to more accurately and efficiently perform spin dynamics simulation. The FME provides a quick means to calculate higher order terms, allowing the disentanglement of the stroboscopic observation and effective Hamiltonian that will be useful to describe spin dynamics at all times in solid-state NMR and understand different synchronized or non-synchronized experiments. The FME offers a simple way to handle multiple incommensurate frequencies and thus open perspectives to deal with multi-mode Hamiltonian in the Hilbert space. This approach can provide new aspects not present in the AHT and the FT such as recursive expansion scheme in Hilbert space that can facilitate the development of new or improvement of existing pulse sequence.

CONCLUSION

In this book chapter, we presented an overview of the AHT and FLT which have been used extensively in the analysis of various experiments in solid-state NMR and have been suitable for designing sophisticated pulse sequences and understanding of different experiments. We also present two developing and emerging theories in solid-state NMR (FME, FE). The combinations of two or more of the theories described within will provide a framework for treating the time-dependent Hamiltonian in NMR in a way that can be easily extended to both synchronized and several non-synchronized modulations. The intention of writing this book chapter is to help bring the current and future prospective theoretical aspects of spin dynamics in NMR to the attention of the NMR community and lead new interactions between nuclear magnetic resonance experts and other specialists. Additional quantitative work will demonstrate further utility of the Fer expansion and Floquet-Magnus expansion in nuclear magnetic resonance, in spin physics and in many other areas.

ACKNOWLEDGMENTS

The authors acknowledge support from Harvard University, Harvard Medical School, Massachusetts General Hospital, and the National Institute of Health. The research was supported by NIH (R01-HL110241 and T32-EB013180).

REFERENCES

[1] Brown, S. P., and Emsley, L. *Handbook of Spectroscopy*, Wiley.

[2] Bjerring, M., Vosegaard, T., Malmendal, A., and Nielsen, N. C. *Concepts Magn. Reson*. A. 2003, 18A, 111-129.

[3] Brown, S. P., and Spiess, H. W. *Chem. Rev.* 2001, 101, 4125-4155.

[4] Schnell, I., and Spiess, H. W. J. *Magn. Reson.* 2001, 151, 153-227.

[5] Mehring, M. *Principles of High Resolution NMR in Solids*, 2nd ed. (Springer, Berlin) 1983.

[6] Feng, X., Verdegem, P. J. E., Lee, Y. K., Sandström, D., Edén, M., Bovee-Geurts, P., de Grip, W. J., Lugtenburg, J., de Groot, H. J. M., and Levitt, M. H. J. *Am. Chem. Soc.* 1997, 119, 6853-6857.

[7] Marassi, F. M., and Opella, S. J. *Curr. Opin. Struct. Biol.* 1998, 8, 640-648.

[8] Smith, S. O., Song, D., Shekar, S., Groesbeek, M., Ziliox, M., and Aimoto, S. *Biochemistry* 2001, 40, 6553-6558.

[9] Cross, T. A., and Quine, J. R. *Concepts Magn. Reson.* 2000, 12, 55-70.

[10] Bjerring, M., Vosegaard, T., Malmendal, A., and Nielsen, N. C. *Concepts Magn. Reson.* Part A. 2003, 18A (2), 111-129.

[11] Mananga, E. S. Ph. D thesis, The City University of New York, 2005.

[12] Kerres, J.A. J. *Membr. Sci.* 2001,185, 29-39.

[13] Bendler, J. T., Fontanella, J. J., Shlesinger, M. F., and Wintersgill, M. C. *Electrochim. Acta.* 2001, 46, 1615-1621.

[14] Jayakody, J. R. P., Stallworth, P. E., Mananga, E. S., Farrington-Zapata, J., and Greenbaum, S. G. J. *Phys. Chem.* B. 2004,108, 4260-4262.

[15] Jayakody, J. R. P., Khalfan, A., Mananga, E.S., Greenbaum, S.G., Dang, T.D., and Mantz, R. *Journal of Power Sources* 2006, 156, 195–199.

[16] Charpentier, T. Habilitation à diriger les recherches, Université Pierre et Marie Curie, Paris VI 2007.

[17] Clark, T. M., Grandinetti, P. J., Florian, P., and Stebbins, J. F. *Phys. Rev.* B, 2004, 70, 064202-1-8.

[18] Levitt, M. H. J. *Chem. Phys.* 2008, 128, 052205-1-25.

[19] Veshtort, M., and Griffin, R. G. J. Magn. Reson. 2006, 178, 248-282.

[20] Bak, M., Rasmussen, J. T., and Nielsen, N. C. J. *Magn. Reson.* 2000, 147, 296-330.

[21] Andrew, E. R., Bradbury, A., and Eades, R. G. *Nature* 1958, 182, 1659-1659.

[22] Andrew, E. R., Bradbury, A., and Eades, R. G. *Nature* (London) 1959, 183, 1802-1803.

[23] Lowe, I. J. *Phys. Rev. Lett.* 1959, 2, 285-287.

[24] Schaefer J., and Stejskal, E. O. J. *Am. Chem. Soc.* 1976, 98, 1031-1032.

[25] Tycko, R. *Annual Review of Physical Chemistry* 2001, 52, 575-606.

[26] Pines, A., Waugh, J. S., and Gibby, M. *J. Chem. Phys.* 1972, 56, 1776-1777.

[27] Hartmann, S. R., and Hahn, E. L. *Phys. Rev*. 1962, 128, 2042-2053.
[28] Tycko, R. J. *Chem. Phys*. 1993, 98, 932-943.
[29] Eden, M., and Levitt, M. H. J. *Chem. Phys*. 1999, 111, 1511-1519.
[30] Levitt, M. H., Raleigh, D. P., Creuzet, F., and Griffin, R. G. J. *Chem. Phys*. 1990, 92, 6347-6364.
[31] Tycko, R. J. *Chem. Phys*. 2008, 128, 052101.
[32] Nielsen, N. C., Vosegaard, T., and Malmendal, A. *Modern Magnetic Resonance*, Graham A. Webb (ed.), Springer, 2008, 679-687.
[33] Haeberlen, U., and Waugh, J. S. *Phys. Rev*. 1968, 175, 453-467.
[34] Ernst, R. R., Bodenhausen, G., and Wokaun, A. Clarendon, Oxford, 1987. Floquet, M. G. *Ann. Econ. Norm. Suppl*. 1883, 12, 47-88.
[35] Shirley, J. H. *Phys. Rev*. B. 1965, 138, 979-987.
[36] Maricq, M. M. *Phys. Rev*. B. 1982, 25, 6622-6632.
[37] Zur, Y., Levitt, M. H., and Vega, S. J. *Chem. Phys*. 1983, 78, 5293-5298.
[38] Madhu, P. K., and Kurur, N. D. *Chem. Phys. Lett*. 2006, 418, 235-238.
[39] Fer, F. Bull. *Classe Sci. Acad. R. Belg*. 1958, 44, 818-829.
[40] Mananga, E. S., and Charpentier, T. J. *Chem. Phys*. 2011, 135, 044109-1-11.
[41] Blanes, S., Casas, F., Oteo, J. A., and Ros, J. *Physics Reports* 2009, 470, 151-238.
[42] Mananga, E. S. *Solid State Nucl. Magn. Reson*, 2013, 55-56, 63-72.
[43] Mananga, E. S. *Solid State Nucl. Magn. Reson*, 2013, 55-56, 54-62.
[44] Mananga, E. S., Reid, A. E., and Charpentier, T. *Solid State Nucl. Magn. Reson*. 2011, 41, 32-47.
[45] Mananga, E. S., and Reid, A. E. *Mol. Phys*. 2013, 111, 2, 243-257.
[46] Evans, W. *Ann. Phys*. 1968, 48, 72-93.
[47] Burum, D. P. *Phys. Rev*. B 1981, 24, 3684-3692.
[48] Magnus, W. Commun. *Pure Appl. Math*. 1954, 7, 649-673.
[49] Casas, F., Oteo, J. A., and Ros, J. J. *Phys. A: Math. Gen*. 2001, 34, 3379-3388.
[50] Blanes, S., Casas, F., Oteo, J. A., and Ros, J. J. *Phys. A: Math. Gen*. 1998, 22, 259-268.
[51] Klarsfeld, S., and Oteo, J. A. *Phys. Rev*. A 1989, 39, 3270-3273.
[52] Klarsfeld, S., and Oteo, J. A. *Phys. Lett*. A 1989, 142, 393-397.
[53] Casas, F. J. Phys. A: *Math. Theor*. 2007, 40, 15001-15017.
[54] Blanes, S., Casas, F., and Ros, *J. BIT* 2000, 40, 434-450.
[55] Blanes, S., Casas, F. and Ros, *J. BIT* 2002, 42, 262-284.
[56] Oteo, J. A., and Ros, J. *J. Math. Phys*. 2000, 41, 3268-3277.

[57] Iserles, A., Norsett, S. P., and Rasmussen, A. F. *Appl. Numer. Math.* 2001, 39, 379-401.

[58] Oteo, J. A., and Ros, J. *J. Phys. A: Math. Gen.* 1991, 24, 5751-5762.

[59] Moan, P. C., and Niesen, *J. Found. Comput. Math.* 2008, 8, 291-301.

[60] Casas, F., and Iserles, A. J. *Phys. A: Math. Gen.* 2006, 39, 5445-5461.

[61] Goldman, M., Grandinetti, P. J., Llor, A., Olejniczak, Z., Sachleben, J. R., and Zwanziger, J. W. J. *Chem. Phys.* 1992, 97, 8947-8960.

[62] Mananga, E. S., Roopchand, R., Rumala, Y. S., and Boutis, G. S. J. *Magn. Reson.* 2007, 185, 28-37.

[63] Mananga, E. S., Rumala, Y. S., and Boutis, G. S. *J. Magn. Reson.* 2006, 181, 296-303.

[64] Mananga, E. S., Hsu, C. D., Ishmael, S., Islam, T., and Boutis, G. S. *J. Magn. Reson.* 2008, 193, 10-22.

[65] Rhim, W. K., Pines, A., and Waugh, J. S. *Phys. Rev.* B. 1971, 3, 684-696.

[66] Matsui, S. *Chem. Phys. Lett.* 1991, 179, 187-190.

[67] Hepp, M. A., and Miller, J. B. J. *Magn. Reson.* A. 1994, 111, 62-69.

[68] Eden, M., and Levitt, M. H. J. *Chem. Phys.* 1999, 111, 1511-1519.

[69] Tycko, R., and Dabbagh, G. *Chem. Phys. Lett.* 1990, 173, 461-465.

[70] Nielsen, N. C., Bildsoe, H., Jakobsen, H. J., and Levitt, M. H. J. *Chem. Phys.* 1994, 101, 1805-1812.

[71] Ishii, Y. J. *Chem. Phys.* 2001, 114, 8473-8483.

[72] Bennett, A. E., Rienstra, C. M., Griffiths, J. M., Zhen, W. G., Lansbury, P. T., and Griffin, R. G. J. *Chem. Phys.* 1998, 108, 9463-9479.

[73] Chan, J. C., and Tycko, R. J. *Chem. Phys.* 2004, 120, 8349-8352.

[74] Costa, P. R., Sun, B. Q., and Griffin, R. G. *J. Am. Chem. Soc.* 1997,119, 10821-10830.

[75] Ladizhansky, V., and Griffin, R. G. *J. Am. Chem. Soc.* 2004, 126, 948-958.

[76] Nomura, K., Takegoshi, K., Terao, T., Uchida, K., and Kainosho, M. *J. Am. Chem. Soc.* 1999, 121, 4064-4065.

[77] Nomura, K., Takegoshi, K., Terao, T., Uchida, K., and Kainosho, M. J. *Biomol. NMR*. 2000, 17, 111-123.

[78] Paravastu, A. K., and Tycko, R. J. *Chem. Phys.* 2006, 124,194303-1-15.

[79] Gullion, T., and Vega, S. *Chem. Phys. Lett.* 1992, 194, 423-428.

[80] Bennett, A. E., Rienstra, C. M., Griffiths, J. M., Zhen, W. G., Lansbury, P. T., and Griffin, R. G. J. *Chem. Phys.* 1998, 108, 9463-9479.

[81] Brinkmann, A., Eden, M., and Levitt, M. H. J. *Chem. Phys.* 2000, 112, 8539-8554.

[82] Hohwy, M., Jakobsen, H. J., Eden, M., Levitt, M. H., and Nielsen, N. C. *J. Chem. Phys.* 1998, 108, 2686-2694.
[83] Brinkmann, A., and Levitt, M. H. J. *Chem. Phys.* 2001, 115, 357-384.
[84] Carravetta, M., Eden, M., Zhao, X., Brinkmann, A., and Levitt, M. H. *Chem. Phys. Lett.* 2000, 321, 205-215.
[85] Tycko, R. J. *Chem. Phys.* 2007, 126, 064506(1-9).
[86] Tycko, R. Lecture notes, *First Winter School on Biomolecular SSNMR,* Vermont, Jan. 2008.
[87] Sanctuary, B. C., and Cole, H. B. R. J. *Magn. Reson.* 1987, 71, 106-115.
[88] Levitt, M. H. J. *Magn. Reson.* 1982, 48, 234-264.
[89] Tycko, R., Cho, H. M., Schneider, E., and Pines, A. J. *Magn. Reson.* 1985, 61, 90-101.
[90] Ramamoorthy, A., Narasimhan, P. T. *J. Mol. Struct.* 1989, 192, 333-344.
[91] Leskes, M., Madhu, P. K., Vega, S. *Prog. Nucl. Magn. Reson. Spectrosc.* 2010, 57, 345-380.
[92] Scholz, I., Van Beek, J. D., Ernst, M. *Solid State Nucl. Magn. Reson.* 2010, 37, 39-59.
[93] Maricq, M. M. J. *Chem. Phys.* 1986, 85, 5167-5174.
[94] Brown, S. P. *Solid State Nucl. Magn. Reson.* 2012, 41, 1-27.
[95] Brown, S. P., and Spiess, H. W. *Chem. Rev.* 2001, 101, 4125-4155.
[96] Brown, S. P. *Prog. Nucl. Magn. Reson. Spectrosc.* 2007, 50, 199-251.
[97] Schnell, I., and Spiess, H. W. *J. Magn. Reson.* 2001,151, 153-227.
[98] Filip, C., Filip, X., Demco, D. E., and Hafner, S. *Molec. Phys.* 1997, 92, 757-771.
[99] Friedrich, U., Schnell, I., Brown, S. P., Lupulescu, A., Demco, D. E., and Spiess, H. W. *Molec. Phys.* 1998, 95, 1209-1227.
[100] Levante, T. O., Baldus. M., Meier, B. H., and Ernst, R. R. J. M. *Molec. Phys.* 1995, 86, 1195-1212.
[101] Boender, G. J., Vega, S., de Groot, H. J. M. J. *Chem. Phys.* 2000, 112, 1096-1106.
[102] Ramachandran, R., Bajaj, V. S., and Griffin, R. G. J. *Chem. Phys.* 2005, 122, 164503(1-14).
[103] Scholz, I., Meier, B. H., and Ernst, M. J. *Chem. Phys.* 2007, 127, 204504(1-13).
[104] Ramesh, R., and Krishnan, M. S. J. *Chem. Phys.* 2001, 114, 5967-5973.
[105] Kotani, S., Chaos, *Solitons and Fractals* 1997, 8, 1817-1854.
[106] Boender, G. J., Vega, S., de Groot, H. J. M. *Molec. Phys.* 1998, 95, 921-934.

[107] Madsen, D. K., Xiao, Y., and Eno, L. J. *Chem. Phys.* 1996, 104, 1344-1349.
[108] Ernst, M., Samoson, A., and Meier, B. H. J. *Chem. Phys.* 2005, 123, 064102(1-10).
[109] Ramachandran, R., and Griffin, R. G. J. *Chem. Phys.* 2005, 122, 164502(1-10).
[110] Ding, S., and McDowell, C. A. *Chem. Phys. Lett.* 1998, 288, 230-234.
[111] Leskes, M., Akbey, U., Oschkinat, H., van Rossum, B. J., and Vega, S. *J. Magn. Reson.* 2011, 209, 207-219.
[112] Buishvili, L. L., Volzhan, E. B., and Menabde, M. G. *Theor. Math. Phys.* 1981, 46, 166-173.
[113] Buishvili, L. L., Gumberidze, M., Menabde, M. G. and Sumenko, O. V. *Theor. Math. Phys.* 1985, 64, 965-969.
[114] Vega, S. *Second Winter school on Biomolecular Solid-State NMR Stowe,* Vermont, Jan. 24-29, 2010.
[115] Boender, G. J., Vega, S., and De Groot, H. J. M. *Mol. Phys.* 1998, 95, 921-934.
[116] Ramachandran, R., and Griffin, R. G. J. *Chem. Phys.* 2005,122, 164502(1-10).
[117] Sanctuary, B. C. J. *Chem. Phys.* 1976, 64, 4352-4361.
[118] Sanctuary, B. C. *Mol. Phys.* 1983, 48, 1155-1176.
[119] Abragam, A. *The Principles of Nuclear Magnetism* (Clarendon, Oxford) 1961.
[120] Maricq, M. M., and Waugh, J. S. J. *Chem. Phys.* 1979, 70, 3300-3316.
[121] Tycko, R., and Dabbagh, G. Chem. *Phys. Lett.* 1990, 173, 461-465.
[122] Sanctuary, B. C., and Halstead, T. K. *Adv. Magn. Opt. Reson.* 1990, 15, 79-161.
[123] Edmonds, A. R. *Angular Momentum in Quantum Mechanics*, 3rd ed. Princeton University Press, Princeton, NJ, 1974.
[124] Silver, B. L. *Irreducible Tensor Methods: An Introduction for Chemists* Academic, New York, 1976.
[125] Haeberlen, U. *Advances in Magnetic Resonance*, Suppl. 1 (New York: Academic Press) 1976.
[126] Ernst, M. J. *Magn. Reson.* 2003, 162, 1-34.
[127] Wilcox, R. M. J. *Math. Phys.* 1967, 8, 962-982.
[128] Zanna, A. *Appl. Numer. Math.* 2001, 39, 435-459.
[129] Suying, Z., and Zichen, D. *Mechanics Research Communications* 2004, 31, 213-219.

[130] Casa, F., Oteo, J. A., and Ros, J. *J. Phys. A: Math. Gen*. 1991, 24, 4037-4046.

[131] Cacas, F. *Numer. Math.* 1996, 74, 283-303.

[132] Klarsfeld, S., and Oteo, J. A. J. *Phys. A: Math. Gen.* 1989, 22, 2687-2694.

[133] Hafner, S., and Demco, D. E. *Solid State Nucl. Magn. Reson*. 2002, 22, 247-274.

[134] Mehring, M. *High resolution NMR Spectroscopy in Solids,* Springer-Verlag, Germany, 1976.

[135] Schmidt, A., and Vega, S. J. *Chem. Phys*. 1992, 96, 2655-2680.

[136] Llor, A. *Chem. Phys. Lett*. 1992, 199, 383-390.

[137] Feike, M., Demco, D. E., Graf. R., Gottwald, J., Hafner, S., and Spiess, H. W. *J. Magn. Reson*. A. 1996, 122, 214-221.

[138] Lee, Y. K., Kurur, N. D., Helmle, M., Johannessen, O. G., Nielsen, N. C., and Levitt, M. H. *Chem. Phys. Lett*. 1995, 242, 304-309.

[139] Spiess, H. W. NMR, *Basic Principles and Progress*, Springer-Verlag, Berlin, Vol. 15, 1978.

In: Nuclear Magnetic Resonance (NMR) ISBN: 978-1-63117-768-2
Editor: Doddapuneni Krishna Rao © 2014 Nova Science Publishers, Inc.

Chapter 7

A QUANTITATIVE ^{1}H HR-MAS STUDY OF THE METABOLIC PROFILE OF THE PGI CHERRY TOMATO OF PACHINO

Carmelo Corsaro,[1] Domenico Mallamace[2,3] and Giacomo Dugo[2]

[1]Dipartimento di Fisica e Scienze della Terra,
Università di Messina, Messina, Italy
[2]Dipartimento di Scienze dell'Ambiente, della Sicurezza,
del Territorio, degli Alimenti e della Salute,
Università di Messina, Messina, Italy
[3]Istituto Zooprofilattico Sperimentale della Sicilia "A. Mirri,"
Palermo, Italy

ABSTRACT

The Italian cherry tomato of Pachino is the first tomato that received the European PGI (Protected Geographical Indication) certification of quality. Its organoleptic and healthy properties are due to different reasons among which are climate, salt water irrigation, cultivation techniques and soil consistency. The well-known reputation of the cherry tomato of Pachino and the relatively high price of the final product originated commercial frauds in the Italian and international markets. As a consequence, great interest has recently been focused on analytical techniques able to verify the geographical origin of a tomato sample, indicating whether or not it originates from the area of Pachino, Sicily (Italy).

In this work we present a quantitative ^{1}H HR-MAS NMR study of the metabolites constituting the PGI cherry tomato of Pachino. This technique allows us to analyze a few samples with a very high resolution. The aim is to develop a protocol allowing a powerful and easy analysis of cherry tomato samples present on the market in order to verify the real origin and detect possible frauds. This would imply the safeguard of the PGI or PDO (Protected Designation of Origin) marks protecting both producers and consumers.

Keywords: HRMAS-NMR, ^{1}H NMR, cherry tomato, metabolite profiling, metabolomics, food science

INTRODUCTION

The tomato (*Solanum lycopersicum*) is an important component of healthy diets, such as the Mediterranean. In fact, it displays a strong antioxidant activity due to the relatively high concentration of lycopene, together with ascorbic acid, vitamin E, flavonoids etc. Tomatoes are consumed either fresh or as a range of processed products and have very distinct quality features [1]; in particular, fresh ones must have acceptable flavour and handling characteristics in order to satisfy the consumers demand and distribution requirements. Furthermore, tomatoes are also used by the food industry as raw materials or flavoring additives for a wide variety of products. For these reasons tomato production is related to the increasing demand of quality by consumers.

The typical Sicilian cherry tomato of Pachino was accredited by the European PGI certificate, a certification that guarantees the geographical origin of a product and specific quality criteria [2, 3]. The motivation for this identification is due to its organoleptic and healthy properties as well as to its sweet taste. The main chemical constituents that are responsible for the particular taste of tomatoes are sugars, organic acids, free amino acids and salts [4]. Therefore, the improvement of both organoleptic and nutritional properties of tomatoes requires identifying, analyzing and interpreting the significance of the different chemical constituents of tomatoes.

Nowadays, food characterization is a very challenging topic because it includes both authenticity and geographical origin determination [5, 6]. Among all possible approaches, metabolomics has shown its great potentiality in the field of living systems, showing how the metabolites content is important for a comprehensive analysis of all metabolic changes upon stimuli [7]. In particular, metabolomics is the scientific discipline able to identify and quantify the different metabolites that determine and characterize the studied bio-system in its particular state [7]. As is well known, metabolism represents the physical and chemical reactions that occur in a living organism (or part thereof), and consists essentially in transformations of matter caused by energetic variations. The metabolic content constitutes a sort of fingerprint for a given bio-system that in principle enables its univocal identification. Many studies took advantage of the multivariate statistical approach to evaluate large sets of information obtained by advanced analytical techniques and to discriminate information dealing with redundancy often present in metabolomic data [8]. Metabolite profiling, sometimes used as synonym of metabolomics, is probably the strategy that offers the widest applicability for metabolic phenotype characterization. The two main metabolic profiling approaches consist of mass-spectrometry (MS) coupled with liquid or gas chromatographic separation (LC–MS and principally GC–MS) and NMR methodologies [9-13]. In comparison with MS-based approaches the NMR technique is less sensitive however, it has important advantages. In fact, NMR spectroscopy is a rapid, non-destructive, high-throughput method for identifying and quantifying plant metabolites [14-17]. NMR is probably the most selective analytical technique available and it may be considered as a universal detector. It is a non-biased method, in which separation and/or derivatization is not required. It is possible to perform the analysis both *in vivo* and on unfractionated or fractionated extracts.

In this work we present an experimental study aimed to characterize and discriminate the metabolic profile of the PGI cherry tomato of Pachino. The production area of the cherry tomato of Pachino includes the entire municipality of Pachino and Portopalo and part of the territories of Noto and Ispica, located in the south east of Sicily (Italy). In this land the cultivation of horticultural products finds ideal soil and climate conditions due to the high exposure to the sun, the salinity of the irrigation water, the soil consistency and the proximity of the sea.

As mentioned, during the last years the need for the geographical characterization of horticultural products increased for both consumers and producers, and several studies essentially based on isotopes, multielement analysis, or separation procedures were applied to different foods [18, 19].

For our purpose, we used the nuclear magnetic resonance (NMR) technique at 700 MHz in the condition known as Magic Angle Spinning

(MAS) in order to canceling the contribution due to the dipolar interactions so obtaining very high resolution NMR spectra even on few amount of sample (in the order of 50 μl).

In principle, proton High Resolution Magic Angle Spinning Nuclear Magnetic Resonance (HR-MAS NMR) allows for simultaneous analysis of polar and apolar components of food samples without requiring separation or extraction steps [20]. In particular, this technique represents a recent and extremely powerful NMR method for the investigation of micro quantities of semi-solid compounds without the need of time-consuming separation procedures. Moreover, the suitability of this technique for the detection of low concentration bioactive molecules has been pointed out [21].

Among different spectroscopic methods, high resolution NMR and above all HR-MAS NMR is a well-suited technique to study complex materials and in particular biological samples such as cells, DNA, proteins, intact tissues, organs, food specimens, membranes, etc. [22-27].

Very recently, this technique was successfully applied also to study the degradation products of ancient paper in order to achieve useful information about production procedures (and origin) and conservation strategies [28].

Thanks to the potential advantages of this methodology in the metabolomic area, several studies on tomatoes have been published in the last years [29-33]. One of them was focused on the application of the HR-MAS NMR technique for metabolic profiling of tomatoes and in particular for tissue differentiation and fruit ripening [30]. The authors find that tryacylglycerols were the major components of seeds. By contrast, signals of fructose, glucose, citric acid, 4-aminobutyric acid (GABA) and the amino acids glutamine and glutamate dominated the NMR spectra of flesh, peel and tomato purée. Another study was focused on the discrimination of triple concentrated tomato paste coming from Italy and China by means of ^{1}H NMR spectroscopy [32]. Citrate content resulted in being the most relevant chemical compound for Chinese and Italian sample differentiation together with aspartate, glutamine, and sugars concentration.

The study here presented concerns the determination of the molar concentration of the major content of metabolites within different samples of cherry tomatoes of Pachino. The outcome was achieved by means of a precise sample preparation and peaks assignment. This latter task was performed by means of literature data [29-33] and of a well-established software package: Chenomx 7.1 (see methods section).

In our study we consider first 14 different kinds of cherry tomato samples of Pachino in order to perform a quantitative study of the molar concentration of metabolites. After that, we include in our analysis 12 different cherry tomato samples of dubious provenience (not Pachino) and only 2 coming directly from Beijing (China) for a multivariate statistical analysis in terms of the Principal Components Analysis (PCA). The PCA analysis is performed to highlight the metabolic differences between different cherry tomato species [30-32, 34].

METHODS

Tomato samples were accurately chosen at the same ripening stage because differences in the ripening stage correspond to differences in the metabolic content [30, 31, 35]. Several samples for each kind of tomato were analyzed in order to increase the statistics. The preparative of tomato samples consists of diluting 6 mg of freeze-dried tomatoes in 100 μl of D_2O (Sigma-AldrichCo) containing 1 mM 2,2-dimethyl-2-silapen-tane-5-sulfonate (DSS) (Sigma-AldrichCo). The DSS acts as an internal standard for the determination of the metabolites concentration providing also a chemical shift reference for the NMR spectra. D_2O is important because it provides a deuterium lock for the NMR spectrometer (necessary for a fine optimization of the magnetic field homogeneity) and it is a good solvent (as well as H_2O) without originating any proton signals. Fifty microliters of each sample were then pipetted into a 4 mm-diameter zirconia sample holder (rotor) with a spherical insert and a Kel-F rotor cap. The NMR experiments were performed by using a Bruker Avance 700 MHz. As said, the experimental configuration used is known as "Magic Angle Spinning" and allows obtaining NMR spectra at very high resolution even on very few amounts of sample (in the order of 50 μl) in liquid and semi liquid phase. By tilting samples of a precise angle (about 54.7°) with respect to the direction of the applied magnetic field, the Hamiltonian term corresponding to dipolar interactions vanishes and NMR peaks become narrower. Furthermore, by spinning the rotor at the magic angle by few thousands of Hertz, line broadening effects due to susceptibility differences within the sample are removed resulting in high resolution quality spectra. The sample temperature was kept at 300 K and controlled by a cold N_2 flow and a heating element, calibrated against the standard CH_3OH reference (4% CH_3OH in CD_3OD) with an accuracy of *0.2 K*. In our case, the rotor was spun at 6000 Hz; the duration of the hard pulse was of 8 μs with a relative attenuation of 3dB; the spectral width was 10 kHz, the acquisition time 2.9s, the points in the time domain 64k, the number of transients 128 and the relaxation time 2s for a total time of about 10 min per experiment. For the reduction of the residual signal of water, we use the standard Bruker presaturation pulse sequence “zgpr”. All spectra were processed (before performing the statistical analysis) by using the routines of the Bruker software Xwinnmr 3.5.

As mentioned above, individual metabolites were identified by means of NMR Suite Professional version 7.1 (Chenomx, Alberta, Canada). This is a well established software package based on a highly sophisticated targeted profiling technology which allows an easy deconvolution of complex NMR spectra into their individual components with good accuracy. The Chenomx reference libraries contain hundreds of fully searchable compound models that are automatically calibrated by adjusting compound line shapes, peak widths, and chemical shifts to better match the sample conditions.

NMR spectra were converted to a format for the multivariate analysis using custom-written ProMetab3.3 software [36] in MATLAB (versionR2009b; The Math Works, Natick, MA). Each spectrum was segmented into 0.005 ppm chemical shift bins between 0.7 and 10.0 ppm, with bins from 4.65 to 4.95 ppm (residual water signal) excluded from all NMR spectra. The area for each segmented region was calculated and normalized to the total area of the spectra. All NMR spectra were generalized by log transformation (with a transformation parameter, $\lambda = 10^{-6}$) to stabilize the variance across the spectral bins and to increase the weightings of the less intense peaks [37]. Data were mean-centered before Principal Components Analysis (PCA) using the Unscrambler X package (version 10.0.1; Camo Software AS, Oslo, NO) and the singular value decomposition (SVD) algorithm was applied to perform a PCA with cross validation. PCA, an unsupervised pattern recognition technique, allowed the differences and similarities between NMR metabolic fingerprints to be visualized in a score plot, where samples that are metabolically similar cluster together. The corresponding PCA loadings plot was used to identify the metabolic basis of the clustering. Representative proton peaks were identified and the Student’s t-test was used to consider the significant metabolic changes between tomato groups (Microsoft Excel).

Discussion

Figure 1 shows a representative ^{1}H HRMAS NMR spectrum of a cherry tomato of Pachino in the region of amino acids. In the figure the assignment of all metabolites is reported. As one can see, the biggest signals belong to citric acid, β-glucose and GABA (4-aminobutyric acid). Most abundant amino acids (see Table I) are glutamine, glutamate, asparagine, aspartate, alanine, arginine, proline and leucine. Moreover, the peak at about 3.3 ppm is assigned to methanol, that at 3.2 ppm to choline and the bands at about 0.9 and 1.3 ppm to saturated and unsaturated fatty acids, respectively. Fatty acids were found to be mostly present in tomato peels and seeds [30].

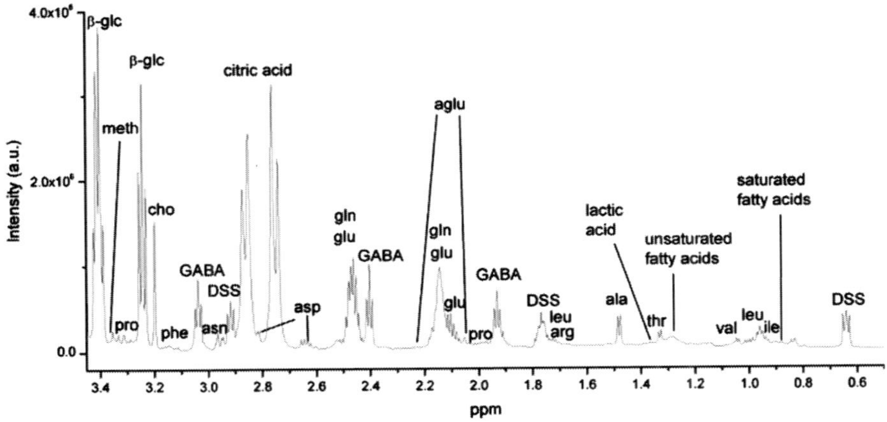

Figure 1. ^{1}H HR-MAS NMR spectrum of the cherry tomato of Pachino in the amino acids region. GABA: 4-aminobutyric acid; aglu: acetylglutamate; ala: alanine; arg: arginine; asn: asparagine; asp: aspartate; β-glc: beta-glucose; cho: choline; gln: glutamine; glu: glutamate; ile: isoleucine; leu: leucine; meth: methanol; phe: phenylalanine; pro: proline; thr: threonine; val: valine.

Figure 2 reports on a different scale the spectral region of sugars, nominally glucose and fructose that are the most abundant substances in tomatoes. In particular, it is also possible to distinguish between β- and α-glucose originating two different signals at 4.62 and 5.22 ppm, respectively. Other relevant metabolites in this spectral region (see Table I) are sucrose, trigonelline, uridine and galactose.

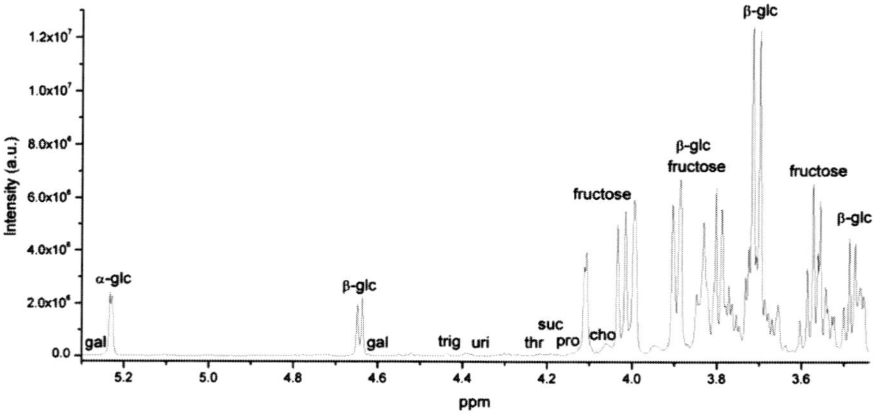

Figure 2. ^{1}H HR-MAS NMR spectrum of the cherry tomato of Pachino in the sugars region. α-glc: alpha-glucose; β-glc: beta-glucose; cho: choline; gal: galactose; pro: proline; suc: sucrose; thr: threonine; trig: trigonelline; uri: uridine.

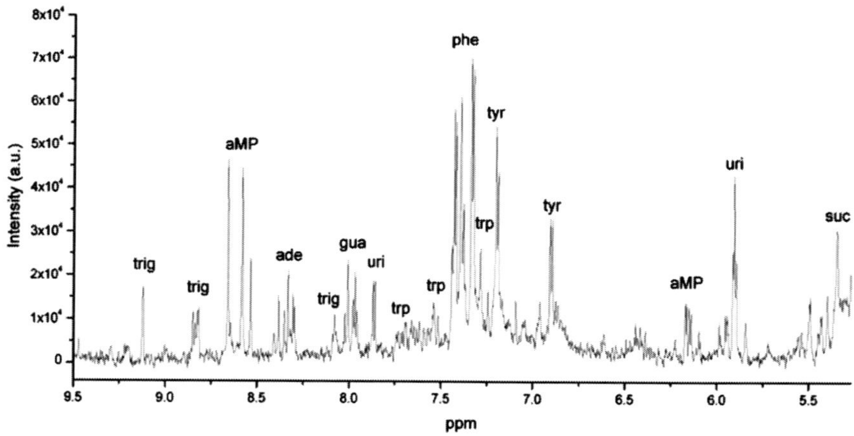

Figure 3. ^{1}H HR-MAS NMR spectrum of the cherry tomato of Pachino in the phenolics regions. ade: adenosine; aMP: adenosine-Mono Phosphate; gua: guanosine; phe: phenylalanine; suc: sucrose; trig: trigonelline; trp: tryptophan; tyr: tyrosine; uri: uridine.

The low-field (highest ppm) region of the spectrum is reported in figure 3 with the corresponding metabolite assignment. In this spectral region the y-scale is about two orders of magnitude smaller than that of sugars region and so the noise signal is bigger. Nevertheless some relevant peaks are easily observed. The most intense were assigned to phenylalanine, tyrosine, adenosine-Mono Phosphate (aMP) and uridine.

The average metabolites concentration, obtained by means of Chenomx software package, is reported in Table 1 together with the corresponding standard deviation. The obtained concentrations can be considered as the fingerprint of the cherry tomato of Pachino metabolic content. However, one has to consider that by using water as solvent, the polar compounds are preferentially visible. Further studies will consider different kinds of solvents to quantify for example also lycopene content.

The listed molar concentrations are consistent and display a relatively small standard deviation; however, they need to be compared with those of different cherry tomato species in order to identify which chemical constituents can be considered relevant for sample differentiation.

Therefore, with the aim to find out those metabolites whose presence and concentration can be considered relevant for sample differentiation, and so to characterize the cherry tomato of Pachino, we performed a multivariate statistical analysis in terms of PCA (see Methods section). In doing so, we considered additional 12 cherry tomatoes of dubious provenience and 2 coming from Beijing (China). Figure 4 shows the score plot of the two principal components (PC1 vs PC2) that demonstrates how this kind of analysis is able to discriminate different tomato species. In particular PC1 well separates, except for one sample, Pachino tomato samples with respect to not Pachino ones.

The loadings corresponding to the Principal Component (PC1), that are able to discriminate between cherry tomatoes of Pachino and not (including those from Beijing), contain information about the different metabolic content (Figures 5 and 6). In particular, Figure 5 shows the ppm range from 0.3 to 5 (amino acids and sugars region) highlighting how Pachino tomatoes have a higher (negative values of the loadings) sugar (fructose and glucose) amount and a lower (positive values of the loadings) fatty acids (both saturated and unsaturated) content. Other visible metabolic differences concern alanine, methanol and acetylglutamate content, which is lower in Pachino cherry tomatoes. On the contrary, GABA, glutamine, glutamate and aspartate concentration is higher in Pachino cherry tomatoes. It is noteworthy that GABA is the main inhibitory neurotransmitter in mammals, responsible in the regulation of neuronal excitability throughout the nervous system.

Table 1. Average metabolite molar concentration of cherry tomatoes of Pachino with the corresponding standard deviation

Metabolite	Average concentration (mM)	Standard deviation (mM)
GABA	0.895	0.189
Acetylglutamate	0.041	0.008
Adenosine	0.045	0.011
AdenosineMP	0.035	0.009
Alanine	0.259	0.036
Arginine	0.249	0.066
Asparagine	0.405	0.084
Aspartate	0.807	0.173
Choline	0.25	0.04
Citric acid	5.69	0.64
Fructose	19.41	1.14
Galactose	0.267	0.081
Glucose	13.44	1.09
Glutamate	1.26	0.21
Glutamine	1.597	0.283
Guanosine	0.033	0.009
Isoleucine	0.067	0.013
Lactate	0.127	0.051
Leucine	0.112	0.026
Methanol	0.614	0.082
Phenylalanine	0.092	0.014
Proline	0.265	0.071
Sucrose	0.038	0.008
Threonine	0.248	0.048
Trigonelline	0.051	0.011
Tryptophan	0.068	0.013
Tyrosine	0.058	0.012
Uridine	0.071	0.016
Valine	0.099	0.017

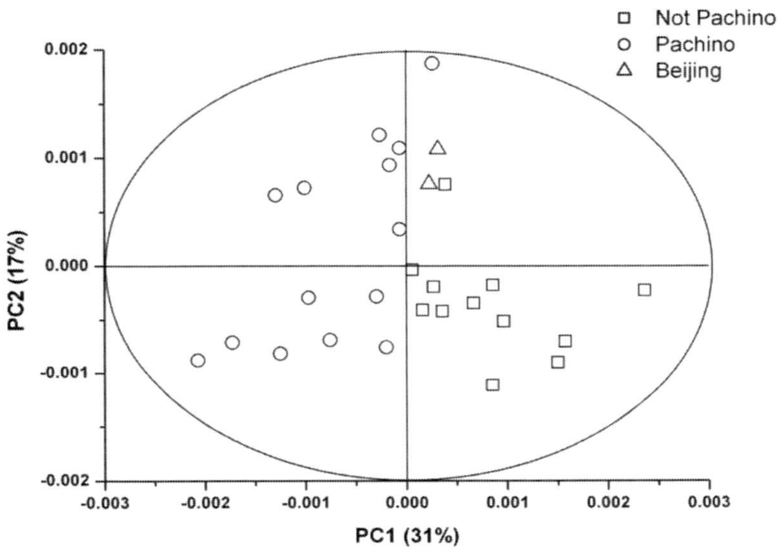

Figure 4. Score Plot of the two principal components of the PCA analysis. The PC1 component is able to discriminate between Pachino and not Pachino cherry tomatoes.

A singular case is that of citric acid, whose peaks are present in both positive and negative side of the PC1 loadings (Figure 5) but they are shifted because of pH. In fact, since we would like to consider also acidic differences between our samples we did not make use of any buffered solution. Citric acid has three carboxylate moieties in its structure, indeed its chemical shift depends on the pH of the solution [38]. In particular lower pH values correspond to higher chemical shift. By looking at Figure 5, one can infer that cherry tomatoes of Pachino are less acidic because the citric acid peaks appear at lower ppm.

The two peaks at about 3.6 and 4.3 ppm, which appear majorly in not Pachino samples, should come from adenosine and malic acid, respectively [39].

Figure 6 reports the PC1 loadings plot in the ppm range from 5.25 to 8.75 (phenolics and aromatics region). As above mentioned, even if the NMR signals in this region are the lowest and mostly affected by noise (see the difference between the y-scale of Figure 2 and Figure 3) several interesting outcomes can be drawn. Guanosine content is lower in Pachino cherry tomatoes whereas trigonelline, tryptophan and tyrosine concentration is higher in Pachino cherry tomatoes. Interestingly, phenylalanine and uridine contents are not discriminant; in fact the molar concentration is about the same in Pachino and not Pachino cherry tomatoes.

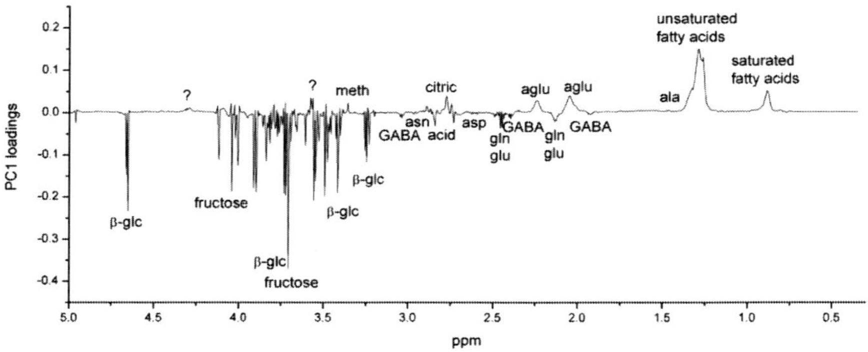

Figure 5. Loadings of the first Principal Component (PC1) corresponding to the performed statistical multivariate analysis in the range from 0.25 to 5.0 ppm.

Finally, having in hands both the metabolites concentration and the loadings results coming from the multivariate statistical analysis, we can identify those metabolites whose concentration (being statistically significant with a p-value less than 0.05) can determine the observed differences.

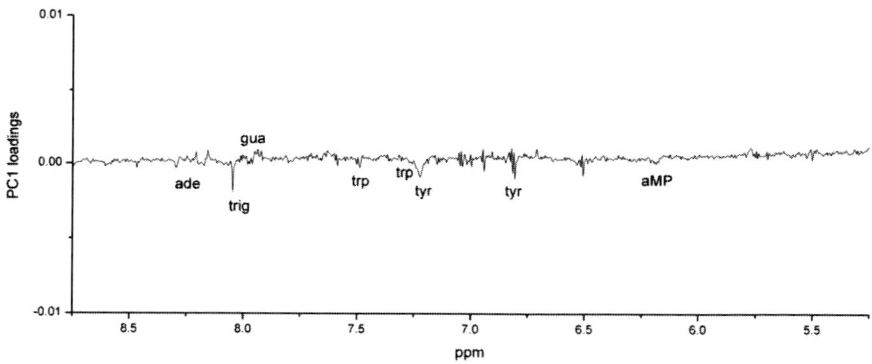

Figure 6. Loadings of the first Principal Component (PC1) corresponding to the performed statistical multivariate analysis in the range from 5.25 to 8.75 ppm.

In Table 2 we report the average concentration of metabolites which are responsible for the revealed metabolic differences that are statistically significant. We have included the two Beijing tomatoes within the not Pachino specie. Our aim is to provide an effective concentration range of those metabolites that characterize the cherry tomato of Pachino peculiar properties. In doing so, it should be possible to distinguish between cherry tomatoes originating within Pachino territory or not.

Table 2. Comparison between the statistically significant average metabolite molar concentration of cherry tomatoes of Pachino and of not Pachino

Metabolite	Average concentration in Pachino cherry tomatoes (mM)	Average concentration in not Pachino cherry tomatoes (mM)
GABA	**0.895**	0.341
Alanine	0.259	**0.543**
Aspartate	**0.807**	0.497
Fructose	**19.41**	14.56
Glucose	**13.44**	10.76
Glutamate	**1.26**	0.78
Glutamine	**1.597**	1.045
Guanosine	0.033	**0.098**
Methanol	0.614	**1.348**
Trigonelline	**0.051**	0.023
Tryptophan	**0.068**	0.032
Tyrosine	**0.058**	0.022

Higher concentrations are reported in bold

CONCLUSION

The reported result allows appreciating the enormous potential and versatility of the technique of investigation that can highlight even the smallest differences in the spectral characteristics of different cherry tomato species. Our study allows quantifying the concentrations of metabolites that generate the observed difference between the considered species. For example, cherry tomatoes of Pachino have a higher concentration of sugars, essential amino acids and GABA. On the other hand they have a smaller amount of metabolites such as fatty acids and methanol, whose excess is not tolerated by the human body.

In conclusion, we show how ^{1}H HR-MAS NMR, coupled with multivariate statistical analysis allows determining the main elements of differentiation concerning the metabolic profile of the cherry tomato of Pachino. The aim is to analyze, easily and efficiently, a tomato sample present on the market in order to verify the real origin and detect possible frauds. This implies the safeguard of PGI or PDO marks [2], protecting both producers and consumers.

ACKNOWLEDGMENT

C. C. thanks the Fondazione Frisone (Catania, Italy) for its support.

REFERENCES

[1] Capanoglu, E.; Beekwilder, J.; Boyacioglu, D.; Hall, R.; De Vos, R. J. *Agric. Food Chem.* 2008, 56, 964-973.
[2] De la Guardia, M.; Illueca, A. G. Food Protected Designation of Origin: Methodologies and Applications, Volume 60 of Comprehensive Analytical Chemistry, Elsevier Science, 2013.
[3] Savorani, F.; Capozzi, F.; Engelsen, S. B.; Dell'Abate, M. T.; Sequi, P. Pomodoro di Pachino: An Authentication Study Using 1H-NMR and Chemometrics – Protecting its P.G.I. European Certification, Magnetic Resonance in Food Science: Challenges in a Changing World, *Royal Society of Chemistry*: Cambridge, UK, 2009, pp. 158 – 166.
[4] Salles, C.; Nicklaus, S.; Septier, C. *Food Chem.* 2003, 81, 395–402.
[5] Luykx, D. M. A. M.; van Ruth, S. M. *Food Chem.* 2008, 107, 897–911.
[6] Ashurst, P. R.; Dennis, M. J. *Analytical Methods Of Food Authentication*, Springer, 1998.
[7] Lindon, J. C.; Nicholson, J. K.; Holmes E. *The Handbook of Metabonomics and Metabolomics,* Elsevier, 2007.
[8] Trygg, J.; Holmes, E.; Lundstedt, T. J. *Proteome Res.* 2007, 6, 469–479.
[9] Moco, S.; Bino, R. J.; De Vos, R. C. H.; Vervoort, J. *Trends in Analytical Chemistry* 2007, 26, 855–866.
[10] Mannina, L. et al. *Journal of Agricultural and Food Chemistry* 2003, 51, 120-127.

[11] Saitta, M.; Giuffrida, D.; La Torre, G. L.; Potorti, A. G.; Dugo, G. *Food Chem.* 2009, 117, 451-455.
[12] Belton, P.S. et al. *J. Agric. Food. Chem.* 1996, 44, 1483-1487.
[13] Moco, S.; Forshed, J.; De Vos, R. C. H.; Bino, R. J.; Vervoort, *J. Metabolomics* 2008, 4, 202–215.
[14] Matto, A. K. et al. Plant Physiol. 2006, 142, 1759–1770.
[15] Defernez, M.; Colquhoun, I. *J. Phytochemistry* 2003, 62, 1009-1017.
[16] Malmendal, A. et al. *J. Agric. Food Chem* 2011, 59, 10831–10838.
[17] Krishnan, P.; Kruger, N. J.; Ratcliffe, R. G. *Journal of Experimental Botany* 2005, 56, 255-265.
[18] Kelly, S.; Heaton, K.; Hoogewerff, J. *Trends Food Sci. Technol.* 2005, 16, 555–567.
[19] Foerstel, H. *Anal. Bioanal. Chem.* 2007, 388, 541–544.
[20] Shintu, L.; Ziarelli, F.; Caldarelli, S. Magnetic *Resonance in Chemistry* 2004, 42, 396–401.
[21] Miglieta, M. L.; Lamanna, R. Magnetic *Resonance in Chemistry* 2006, 44, 675–685.
[22] Torre, A.; Trischitta, F.; Corsaro, C.; Mallamace, D.; Faggio, C. *Cell Biochemistry and Function* 2013, 31, 489-495.
[23] Beckonert, O. et al. *Nat. Protoc.* 2010; 5, 1019-1032.
[24] Tzika, A. A. et al. Int. J. Mol. Med. 2007; 20, 199-208.
[25] Corsaro, C.; Mallamace, D. *Physica A* 2011, 390, 2904–2908.
[26] Jenkins, J. E.; Holland, G. P.; Yarger, J. L. *Soft Matter* 2012, 8, 1947-1954.
[27] Merkley, N.; Syvitski, R. T. *Journal of Applied Phycology* 2012, 24, 535-540.
[28] Corsaro, C. et al. Sci. Rep. 2013, 3, 2896; DOI:10.1038/srep02896.
[29] Sobolev, A. P.; Segre, A.; Lamanna, R. *Magnetic Resonance in Chemistry* 2003, 41, 237-245.
[30] Pérez, E. M. S. et al. *Food Chemistry* 2010, 122, 877-887.
[31] Le Gall, G. et al. *J. Agric. Food Chem.* 2003, 51, 2447–2456.
[32] Consonni, R.; Cagliani, L. R.; Stocchero, M.; Porretta, S. J. *Agric. Food Chem.* 2009 , 57, 4506–4513.
[33] Deborde, C. et al. *Metabolomics* 2009, 5, 183-198.
[34] Porretta, S. Food Qual. Prefer. 1994, 5, 139–144.
[35] Raffo, A. et al. *J. Agric. Food Chem.* 2002, 50, 6550–6556.
[36] Viant, M. R. *Biophys. Res. Commun.* 2003, 310, 943–948.
[37] Parsons, H. M.; Ludwig, C.; Gunther, U.L.; Viant, M.R. BMC *Bioinformatics* 2007, 8, 234.

[38] Miyataka, H.; Ozaxi, T.; Himeno, S. *Biol. Pharm. Bull.* 2007, 30, 667—670.

[39] Mallamace, D. et al. *Physica A* 2014, http://dx.doi.org/10.1016/j.physa.2013.12.054.

In: Nuclear Magnetic Resonance (NMR)
Editor: Doddapuneni Krishna Rao
ISBN: 978-1-63117-768-2
© 2014 Nova Science Publishers, Inc.

Chapter 8

A AND *B* OCCUPATION RATES OF MIXED $A_{2-x}B_x$ZNCL$_4$ (A = NH$_4$, RB, CS; B = RB, CS; x = 0, 1, AND 2) SINGLE CRYSTALS BY NUCLEAR MAGNETIC RESONANCE

Ae Ran Lim[1,2,*]

[1]Department of Science Education, Jeonju University, Jeonju, Korea

[2]Department of Carbon Fusion Engineering, Jeonju University, Jeonju, Korea

ABSTRACT

A_2ZnCl_4 (A=K, Rb, Cs, and NH_4) crystals have received considerable attention for their nonlinear optical properties. This family of crystals has also attracted the attention of researchers for many decades because of the great diversity of their structural phase transitions. Further, some of these crystals reportedly show incommensurate phases. In this study, mixed $A_{2-x}B_xZnCl_4$ (A = NH_4, Rb, Cs; B = Rb, Cs; x = 0, 1, and 2) single crystals were grown from aqueous solutions using the slow evaporation method. The crystal structures and phase transition temperatures of the $A_{2-x}B_xZnCl_4$ crystals were determined through x-ray diffraction and

* Corresponding author. Tel: +82-(0)63-220-2514, Fax: +82-(0)63-220-2053. E-mail: aeranlim@hanmail.net or arlim@jj.ac.kr.

differential scanning calorimetry, respectively. In addition, the nuclear magnetic resonance (NMR) spectra and spin-lattice relaxation time, T_1 of the *A* and *B* nuclei in the $A_{2-x}B_xZnCl_4$ single crystals were obtained by NMR spectroscopy. The preferred approach to probing the structural properties of the mixed $A_{2-x}B_xZnCl_4$ single crystals was by measuring the NMR spectra and relaxation times, because these relaxation times are expected to be very sensitive to changes in the crystal symmetry. The crystallographic structures of these crystals were investigated by determining the *A* and *B* occupation probabilities of the two sites. The NMR experimental results confirm that A_2ZnCl_4 (*A* = NH_4, Rb, and Cs) contains two crystallographically inequivalent *A* groups, *A*(1) and *A*(2), occupying the A_1 and A_2 sites, and that B_2ZnCl_4 (*B* = NH_4, Rb, and Cs) contains two crystallographically inequivalent *B* ions, *B*(1) and *B*(2), occupying the B_1 and B_2 sites. Both types of *A* ions and *B* ions are found in mixed $A_{2-x}B_xZnCl_4$ crystals; the occupation rates for *A*(1) and *A*(2) and those for *B*(1) and *B*(2) are discussed. These results enable us to compare the structural properties and relaxation mechanisms of pure $(NH_4)_2ZnCl_4$, Rb_2ZnCl_4, and Cs_2ZnCl_4 single crystals.

Keywords: $A_{2-x}B_xZnCl_4$, mixed crystals, $(NH_4)_2ZnCl_4$, Rb_2ZnCl_4, Cs_2ZnCl_4, ferroelectrics, nuclear magnetic resonance, spin-lattice relaxation time, phase transition temperature

1. Introduction

Crystals of the general formula A_2BX_4 have attracted much interest, mainly from solid state physicists, for many years because of the great diversity of their structural phase transitions [1-4]. Here, *A* is an alkali metal, and BX_4 is a tetrahedral group (SO_4, SeO_4, CrO_4, $ZnCl_4$, and so on). The prototype of the crystal structures of this family is that of β-K_2SO_4, which consists of isolated BX_4^{2-} tetrahedra and monovalent A^+ cations placed in two inequivalent cavities. A_2BX_4 crystals exhibit a number of interesting phase transitions and can exhibit phases with ferroelectric or ferroelastic ordering [3, 4]. Incommensurate phases have been discovered in some members of this family of crystals. One A_2BX_4 family, $(NH_4)_2ZnCl_4$, undergoes five phase transitions, including those between phases I and II at 406 K (=T_{C1}) and phases II and III at 364 K (=T_{C2}) [5-8]. In addition, successive phase transitions at 319 K (=T_{C3}), 271 K (=T_{C4}), and 266 K (=T_{C5}) have been reported [1, 9, 10]; the phases involved in these transitions are denoted III, , V, and VI in order of decreasing temperature. Above 406 K, the structure of $(NH_4)_2ZnCl_4$ in the normal phase, phase I, is orthorhombic [5]. Upon cooling, a phase transition occurs at 406 K to an incommensurate phase that is stable down to 364 K [11]. The structure in phase III, between 364 K and 319 K, is orthorhombic [11, 12]. The room temperature phase, phase IV, is antiferroelectric with a pseudo-orthorhombic monoclinic structure [11, 13]. Phase V, between 271 K and 266 K, is characterized by the coexistence of several commensurate modulated phases with different periods of modulation. On the other hand, Rb_2ZnCl_4 undergoes successive phase transitions from a paraelectric phase to an incommensurate phase at 303 K (=T_{C1}), to a ferroelectric commensurate phase at 192 K (=T_{C2}), and to a low-temperature phase at 75 K (=T_{C3}) [14-19]. The high-temperature phase is orthorhombic. The intermediate phase is incommensurate, and the low-temperature commensurate phase exhibits improper ferroelectricity with the lattice parameter tripled along the *a*-axis. This crystal undergoes a second-order phase transition from a paraelectric (N) phase to an incommensurate (IC) phase at about 303 K and a first-order phase transition from the IC phase to ferroelectric commensurate at 192 K [20]. The paraelectric unit cell contains eight Rb ions, which belong to two crystallographically inequivalent sites, Rb(1) and Rb(2). In addition, Cs_2ZnCl_4 does not undergo any structural phase transitions, and there are two types of crystallographically inequivalent sites, Cs(1) and Cs(2), in its unit cell [21, 22].

According to previous reports, ^{1}H nuclear magnetic resonance (NMR) relaxation studies of $(NH_4)_2ZnCl_4$ have mainly investigated the dynamical processes at lower temperatures [23, 24]. The proton spin-lattice relaxation time T_1 of polycrystalline $(NH_4)_2ZnCl_4$ at a Larmor frequency of 10 MHz has been reported; double minima were observed in T_1 [25]. Michel et al. [26] reported the quadrupolar perturbed ^{14}N NMR spectra of the normal, incommensurate, and various commensurate phases of $(NH_4)_2ZnCl_4$ crystals. The effects of grain size in polycrystalline samples of $(NH_4)_2ZnCl_4$ were studied with NMR [27]. The deuteron NMR spectra and relaxation of polycrystalline fully and partly deuterated $(NH_4)_2ZnCl_4$ were recently studied between 300 and 5 K [28]. In addition, the spin-lattice relaxation time of the ^{1}H nuclei in the ammonium ions in $(NH_4)_2ZnCl_4$ crystals has been reported as a function of temperature for a range of temperatures that indicates those of the phase transitions [29]. On the other hand, Rb_2ZnCl_4 was probably most often investigated by NMR. Many researchers reported on the soliton density near T_{C1} (=303 K) using a ^{87}Rb NMR technique [30-37]. Further, Lim et al. [38] reported the Raman process by examining ^{87}Rb spin-lattice relaxation time in a Rb_2ZnCl_4 single crystal as a function of temperature. Further,

Cs_2ZnCl_4, which has no structural phase transition, was studied to obtain details on the dynamics of the $ZnCl_4$ anion by measuring the ^{35}Cl nuclear quadrupole resonance (NQR) frequency and relaxation time [39, 40]. Lim et al. [41] reported two different groups of Cs resonances in NMR experimental results for Cs_2ZnCl_4 crystals, which points to the presence of two types of crystallographically inequivalent cesium nuclei, Cs(1) and Cs(2); the quadrupole coupling constants, e^2qQ/h, and asymmetry parameters, η, of Cs(1) and Cs(2) were then obtained: e^2qQ/h=148 kHz and η=0.11 for the Cs(1) ion, and e^2qQ/h=274 kHz and η=0.66 for the Cs(2) ion. Recently, the molecular dynamics and phase transitions of A_2ZnCl_4 (A=NH_4, Rb, and Cs) single crystals were reported and compared for the three crystals [29].

Kolinsky et al. [42] reported the growth of seven different $Rb_{2-x}Cs_xZnCl_4$ crystals with $0 < x < 2$. They found that the distribution of Cs and Rb among the two cationic sites A_1 and A_2 was not random, and they analyzed the crystallographic structures of the crystals by considering the occupation probabilities of the two sites. Rb exhibits a higher affinity for the smaller cavity, the A_2 site. A_1 and A_2 do not have the same environments. The Rb and Cs sites in $RbCsZnCl_4$ crystals are occupied by one Cs in the A_1 site and one Rb in the A_2 site, according to the x-ray diffraction results of Kolinsky et al. [42].

Information regarding the structure and internal motion of solids can be obtained by the NMR technique. From the spin-lattice relaxation time measurements, it was determined that the slope of the spin-lattice relaxation time plotted as a function of temperature changes abruptly near the phase transition temperature. The interest in this research lies in the links between changes in the spin-lattice relaxation times and the structural phase transitions. The NMR method makes it possible to study the local properties of lattices, and is particularly useful when information on the behavior of individual structural groups is required.

To explain the mechanisms of the successive phase transitions of A_2ZnCl_4 (A = NH_4, Rb, and Cs), it is necessary to examine the NMR spectra and spin-lattice relaxation times of all the phases in detail. The aim of this work is to clarify the structural changes in A_2ZnCl_4 associated with the successive phase transitions. Spin-lattice relaxation measurements for A^+ nuclei in A_2ZnCl_4 were investigated using a pulse NMR spectrometer to obtain further information about the phase transitions and their connection with the A^+ ion. In particular, the role of the alkali-metal ions in the structural phase transitions and molecular motions in these compounds were examined. On the basis of these results, the relaxation mechanisms for these crystals were compared.

Further, mixed $A_{2-x}B_xZnCl_4$ ($A = NH_4$, Rb, Cs; B = Rb, Cs; x = 0, 1, and 2) single crystals were grown from aqueous solutions using the slow evaporation method. The structures and phase transition temperatures of the $NH_4RbZnCl_4$, $NH_4CsZnCl_4$, and $RbCsZnCl_4$ crystals were determined by x-ray diffraction and differential scanning calorimetry (DSC), respectively. In addition, the NMR spectra and spin-lattice relaxation times, T_1, for ^1H, ^{87}Rb, and ^{133}Cs in mixed $NH_4RbZnCl_4$, $NH_4CsZnCl_4$, and $RbCsZnCl_4$ single crystals were determined. To probe the structural properties of the three pure and three mixed single crystals, it is preferable to measure the NMR spectra and relaxation times, because they are likely be very sensitive to changes in the symmetry of the crystals. The NMR technique has been shown to be capable of yielding a quantitative determination of the fractional part of the volume of a crystal. The NH_4, Rb, and Cs occupation rates of the two available sites, A_1 and A_2, were determined from the NMR spectra. This is the first time that the structural properties of $A_{2-x}B_xZnCl_4$ crystals have been investigated, and the environments of the *A* and *B* nuclei are analyzed from these results. This enables us to compare the structural properties and relaxation mechanisms of the pure and mixed crystals. The results of this study should be applicable to other isomorphic systems.

2. Experimental

A_2ZnCl_4 ($A = NH_4$, Rb, and Cs) crystals were obtained by slow evaporation at room temperature of aqueous solutions with the appropriate molar ratios of *A*Cl and $ZnCl_2$. The A_2ZnCl_4 crystals were transparent and colorless with hexagonal shapes. Mixed crystals of $A_{2-x}B_xZnCl_4$ ($A = NH_4$, Rb, Cs; B = Rb, Cs; x = 0, 1 and 2) were prepared by slow evaporation of aqueous solutions containing stoichiometric amounts of *A*Cl, *B*Cl, and $ZnCl_2$. The mixed crystals were also transparent, colorless, and hexagonal shapes.

The structure of the $A_{2-x}B_xZnCl_4$ single crystals at room temperature was determined using an x-ray diffractometer system (Bruker AXS GMBH) at the Korea Basic Science Institute. A colorless hexagonal-shaped crystal was mounted on a Bruker SMART CCD diffractometer equipped with a graphite-monochromated Mo Kα (λ=0.71073 Å) radiation source. Data were collected and integrated at 300 K using SMART (Bruker, 2000) and SAINT-Plus (Bruker, 2001) [43]. In addition, to determine the phase transition temperatures, DSC of the crystals was performed using a DuPont 2010 DSC instrument at a heating rate of 10°C/min.

The NMR signals of the ^{1}H nuclei in the $(NH_4)_2ZnCl_4$, $NH_4RbZnCl_4$, and $NH_4CsZnCl_4$ single crystals were measured using the Varian 200 MHz FT NMR spectrometer at the Korea Basic Science Institute. The static magnetic field was 4.7 T and the central radio frequency was set to $\omega_o/2\pi$=200 MHz for the ^{1}H nucleus. The spin-lattice relaxation times were measured using the saturation recovery pulse sequence, *sat–t–*π/2: the nuclear magnetizations of the ^{1}H nuclei at time *t* after the *sat* pulse, a combination of one hundred π/2 pulses applied at regular intervals, were determined following the π/2 excitation pulse. The width of the π/2 pulse was 2.4 μs for ^{1}H. Further, the NMR signals of the ^{87}Rb nuclei in the Rb_2ZnCl_4, $NH_4RbZnCl_4$, and $RbCsZnCl_4$ single crystals were measured using the Varian 200 MHz FT NMR and Bruker 400 MHz FT NMR spectrometer at the Korea Basic Science Institute. The static magnetic fields were 4.7 T and 9.4 T. The central radio frequency was set to $\omega_o/2\pi$=65.48 MHz and 130.93 MHz for the ^{87}Rb nucleus. The spin-lattice relaxation time was measured by applying the pulse sequences π/2–*t*–π/2 and π–*t*–π/2. The nuclear magnetizations M(*t*) of the ^{87}Rb nuclei at time *t* after the π/2 and π pulses were determined from saturation and inversion recovery sequences following the pulses. The widths of the π/2 and π pulses were 1 μs and 1.65 μs for ^{87}Rb, respectively. On the other hand, the NMR signals of ^{133}Cs in Cs_2ZnCl_4, $NH_4CsZnCl_4$, and $RbCsZnCl_4$ single crystals were measured using the Bruker 400 MHz FT NMR spectrometer at the Korea Basic Science Institute. The static magnetic field was 9.4 T, and the central frequency was set at $\omega_o/2\pi$=52.48 MHz for the ^{133}Cs nucleus. The spin-lattice relaxation time was measured by applying the pulse sequences π/2–*t*–π/2 and π–*t*–π/2. The nuclear magnetizations M(*t*) of the ^{133}Cs nuclei at time *t* after the π/2 and π pulses were determined from saturation and inversion recovery sequences following the pulses. The widths of the π/2 and π pulses were 2.5 μs and 5 μs for ^{133}Cs, respectively. The temperature-dependent NMR measurements were obtained over the temperature range 180–430 K. The samples were maintained at constant temperatures by controlling the nitrogen gas flow and heater current.

3. CRYSTAL STRUCTURE

3.1. $(NH_4)_2ZnCl_4$

Above 406 K, the structure of $(NH_4)_2ZnCl_4$ is orthorhombic with space group D_{2h}^{16} (*Pnma*) and lattice constants a=9.274 Å, b=7.211 Å, and c=12.620

Å [5], as indicated in Table 1. Upon cooling, a phase transition to an incommensurate phase occurs at 406 K, which is stable down to 364 K [11, 44, 45]. Between 364 K and 319 K the structure is orthorhombic, but with a space group C_{2v}^9 (*Pn2₁a*) [11, 12, 46] and lattice constants a′=4a, b′=b, and c′=c. The room temperature phase is antiferroelectric with pseudo-orthorhombic monoclinic structure and space group C_s^2 (*Pa*) [13]. At room temperature the lattice constants are a′=4a, b′=b, c′=c and γ=90.0°, as shown in figure 1. The region between 271 K and 266 K is characterized by the coexistence of ferroelectric and ferroelastic phases [47]. This structure consists of NH_4^+ and $ZnCl_4^{2-}$ tetrahedra. In the unit cell, there are two types of NH_4^+ ions, which are crystallographically inequivalent.

3.2. Rb_2ZnCl_4

Rb_2ZnCl_4 undergoes transitions separating phases I, II, III, and IV, at temperatures defined as T_{C1} (=303 K), T_{C2} (=192 K), and T_{C3} (=75 K), as indicated in Table 2. Phase I in the paraelectric state is centrosymmetrical. The structure is orthorhombic with space group D_{2h}^{16} (*Pmcn*), and the lattice constants are a=9.265 Å, b=7.285 Å, and c=12.730 Å [48, 49]. An incommensurate modulation of the superlattice in phase II occurs along the a-axis, and the commensurate superlattice is recovered in phase III with a period of 3a [50]. Phase IV is also ferroelectric and is monoclinic with space group C_s^4 (*C1c1*). The structure of Rb_2ZnCl_4 is shown in figure 2.

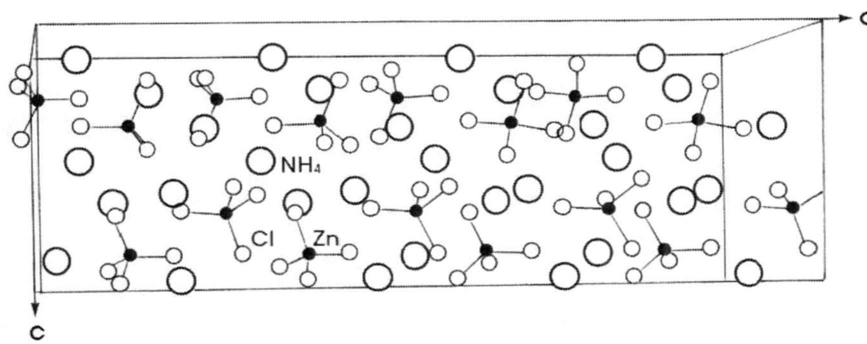

Figure 1. Arrangement of atoms viewed along the b-axis in $(NH_4)_2ZnCl_4$ crystal at room temperature.

Table 1. The phase transition temperature, T_C, the space group, the crystal structure, and the lattice constants (Å) of $(NH_4)_2ZnCl_4$

		T_{C5}		T_{C4}		T_{C3}		T_{C2}		T_{C1}	
		(266 K)		(271 K)		(319 K)		(364 K)		(406 K)	
Phase	VI		V		IV		III		II		I
			Ferroelectric Ferroelastic		Antiferroelectric				Incommensurate		Normal
Space group					*Pa* (C_s^2)		*Pn*2₁*a* (C_{2v}^9)				*Pnma* (D_{2h}^{16})
Structure					Monoclinic		Orthorhombic				Orthorhombic
Lattice constants					a′=4a		a′=4a				a=9.274
					b′=b		b′=b				b=7.211
					c′=c		c′=c				c=12.620

Table 2. The phase transition temperature, T_C, the space group, the crystal structure, and the lattice constants (Å) of Rb_2ZnCl_4

		T_{C3} (75 K)		T_{C2} (192 K)		T_{C1} (303 K)	
Phase	IV		III		II		I
	Ferroelectric		Ferroelectric				Paraelectric
	Commensurate		Commensurate		Incommensurate		Normal
Space group	*C1c1* (C_s^4)		*P2₁cn* (C_{2v}^9)		*Pmcn*		*Pmcn* (D_{2h}^{16})
Structure	Monoclinic		Orthorhombic		Orthorhombic		Orthorhombic
Lattice constants	a=27.439		a′=3a		a=9.221		a=9.265
	b=7.201		b′=b		b=7.253		b=7.285
	c=12.542		c′=c		c=12.646		c=12.730

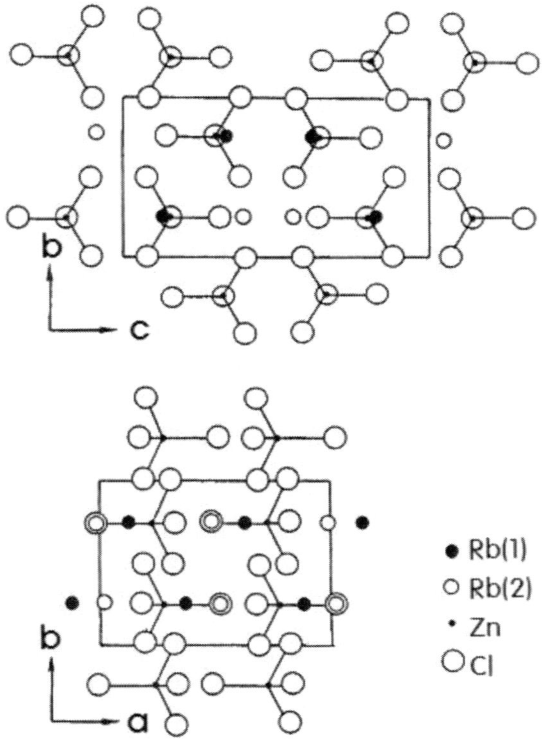

Figure 2. Projections of the atomic positions in Rb_2ZnCl_4 crystal in bc-plane and ab-plane.

The two Rb^+ ions and two of the Cl^- ions from each molecule occupy inequivalent sites on a mirror plane, as does the Zn^{2+} ion [51]. The other two Cl^- ions occupy equivalent sites on opposite sides of the mirror plane [52, 53]. The paraelectric unit cell contains eight Rb ions, which belong to two inequivalent sites, Rb(1) and Rb(2), which lie on the pseudohexagonal and pseudotriad axes, respectively. All the Rb sites have site symmetry *m* and lie in the mirror plane perpendicular to the crystal's b-axis.

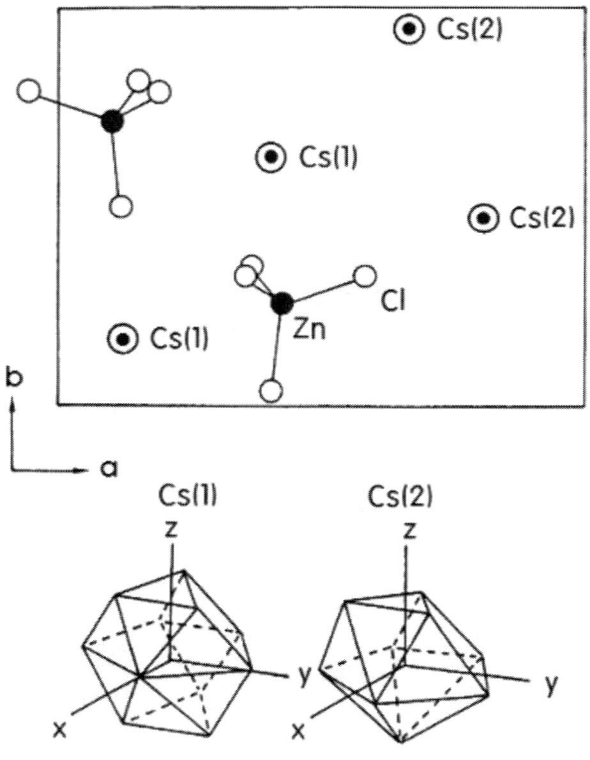

Figure 3. Projection of orthorhombic structure of Cs_2ZnCl_4 crystal along c-axis and surroundings of Cs(1) and Cs(2) ions.

3.3. Cs_2ZnCl_4

The unit cell of the Cs_2ZnCl_4 crystal is orthorhombic and belongs to the space group D_{2h}^{16} (*Pnam*) with four formula units per unit cell. The cell parameters are a=9.76 Å, b=7.40 Å, and c=12.98 Å [54]. Half of the unit cell is shown in figure 3. The unit cell contains four $ZnCl_4^{2-}$ complexes. The local symmetry of the $ZnCl_4^{2-}$ ion is governed by the mirror plane (010) passing through the central ion and two Cl^- ions. However, the local symmetry is close to that of a tetragonal sphenoid. There are two types of crystallographically different Cs^+ ions in the cell [55]. The Cs(1) and Cs(2) ions are surrounded by 11 and 9 chlorine ions, respectively, as shown in figure 3.

4. Experimental Results and Analysis

The saturation recovery traces of the ^1H (I=1/2) nuclei in $(NH_4)_2ZnCl_4$, $NH_4RbZnCl_4$, and $NH_4CsZnCl_4$ can each be represented by a single exponential function. The ^1H spin-lattice relaxation time was determined by fitting the recovery patterns to the following equation [56-61]:

$$[M(\infty) - M(t)] / M(\infty) = \exp(-Wt) \tag{1}$$

where M(t) is the nuclear magnetization at time t, and W is the transition probability corresponding to $\Delta m = \pm 1$. The relaxation time is given by $T_1 = 1/W$.

The recovery laws for the quadrupole relaxation processes of the ^{87}Rb (I=3/2) nuclei in Rb_2ZnCl_4, $NH_4RbZnCl_4$, and $RbCsZnCl_4$ single crystals are described. The saturation recovery function for the central resonance line is [57, 62-65]

$$[M(\infty) - M(t)] / M(\infty) = 0.5\,[\exp(-2W_1t) + \exp(-2W_2t)] \tag{2}$$

where M(t) is the nuclear magnetization corresponding to the central transition at time t after saturation. W_1 and W_2 are the transition probabilities corresponding to $\Delta m = \pm 1$ and $\Delta m = \pm 2$, respectively. Thus, the relaxation time is given by $T_1 = 5 / [2(W_1+4W_2)]$.

When only the central line is excited, the magnetization recovery of the ^{133}Cs (I=7/2) nuclei in Cs_2ZnCl_4, $NH_4CsZnCl_4$, and $RbCsZnCl_4$ crystals does not follow a single exponential function, but can be represented by a combination of four exponential functions. The signal for $W_1 = W_2$ is given by [66-70]

$$[M(\infty) - M(t)] / M(\infty) = 0.048 \exp(-0.476W_1t) + 0.818 \exp(-1.333W_1t) + 0.050 \exp(-2.381W_1t) + 0.084 \exp(-3.810W_1t), \tag{3}$$

where $M(t)$ is the nuclear magnetization corresponding to the central transition at time t after saturation, and W_1 and W_2 are the transition probabilities corresponding to $\Delta m = \pm 1$ and $\Delta m = \pm 2$, respectively. Again the return to equilibrium is characterized by four relaxation times, but because of the much larger coefficient of the second exponential term, the growth is exponential in nature most of the time with a relaxation rate equal to 1.333 W_1.

4.1. ^{1}H NMR Spectrum and Spin-lattice Relaxation Time in $(NH_4)_2ZnCl_4$ Crystals

The NMR spectrum of ^{1}H (I=1/2) in $(NH_4)_2ZnCl_4$ crystals at room temperature is shown in figure 4. The ^{1}H NMR lines due to the crystallographically inequivalent $NH_4(1)$ and $NH_4(2)$ sites overlap. The $NH_4(1)$ and $NH_4(2)$ sites are represented as solid and open circles, respectively, in figure 4. The ^{1}H spectrum is displaced by a chemical shift to the higher frequency side relative to the reference signal obtained for the ^{1}H line from an aqueous solution of H_2O. The intensities of the resonance line with strong intensity (■ in figure 4) depend strongly on temperature, and the intensities of those with low intensity are not easy to measure. For the signals with strong intensity as shown in figure 5, the change in the intensity near T_{C2} (=364 K) is related to the phase transition. The other phase transitions are not evident in the ^{1}H intensity results.

The magnetization decay of the ^{1}H nuclei in $NH_4(1)$ ions with strong intensity (■) in figure 4 was measured for several delay times. Here, the crystallographic b-axis was oriented parallel to the static magnetic field. The lower-intensity magnetization decay for $NH_4(2)$ ions is not easy to measure, and the proton relaxation time was measured for $NH_4(1)$ ions. Figure 6 shows the recovery traces of ^{1}H as a single exponential function at 200, 300, and 400 K. The slopes of the recovery traces are different at each temperature. The recovery data for the temperatures investigated here can all be fitted with equation (1), as shown by the solid lines in figure 6. This figure exhibits the large variation in the proton dynamics of the N-H bond networks.

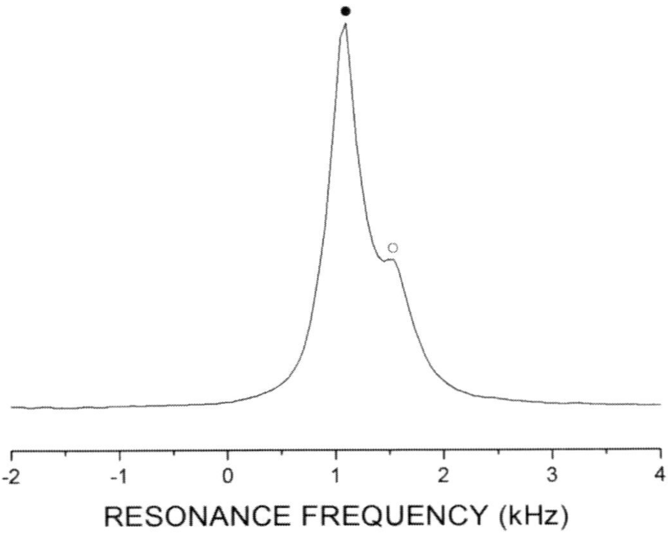

Figure 4. ^1H NMR spectrum of crystallographically inequivalent NH_4 ions in $(NH_4)_2ZnCl_4$ crystal [●: $NH_4(1)$ and ○: $NH_4(2)$].

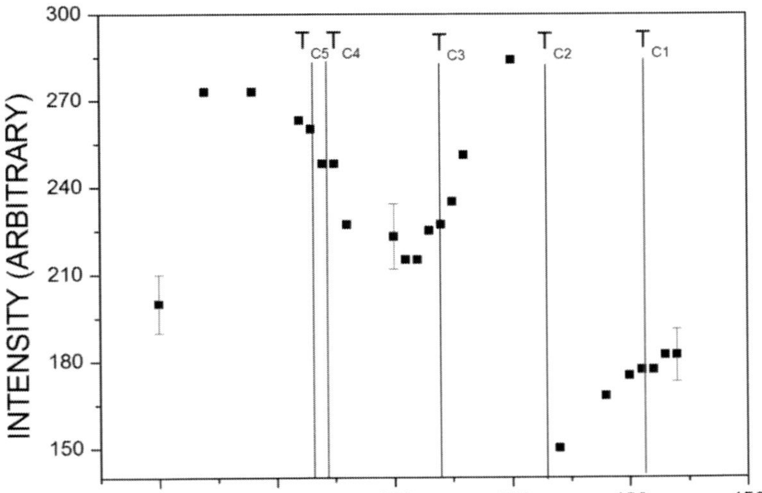

Figure 5. The intensity of ^1H NMR in $(NH_4)_2ZnCl_4$ crystal as a function of temperature.

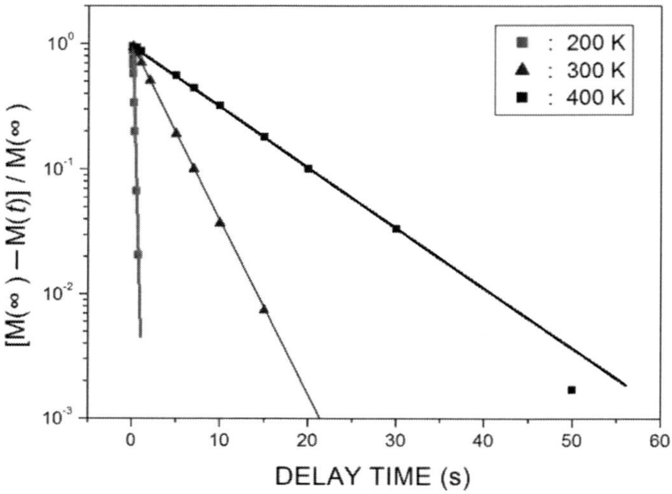

Figure 6. The saturation recovery behaviors of ^{1}H in $(NH_4)_2ZnCl_4$ crystal as a function of delay time at 200 K, 300 K, and 400 K. The solid lines are fits with the function of equation (1).

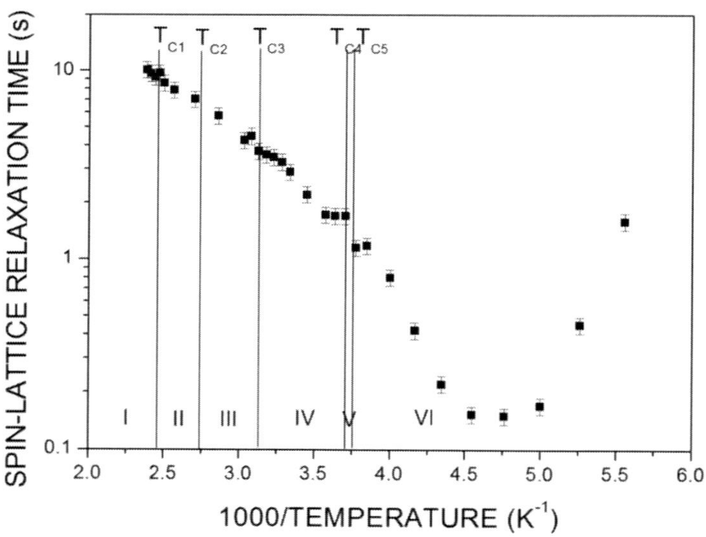

Figure 7. ^{1}H spin-lattice relaxation time in $(NH_4)_2ZnCl_4$ crystal as a function of inverse temperature.

From the slopes of the recovery traces, the ^{1}H spin-lattice relaxation time in $(NH_4)_2ZnCl_4$ was obtained as a function of the inverse temperature, as shown in figure 7. The proton T_1 passes through a small discontinuity near T_{C4} (=271 K), and its slope changes at 271 K, which is probably due to the phase transition from the ferroelectric (or ferroelastic) phase to the antiferroelectric phase. In the high temperature region, T_1 increases monotonically with temperature. T_1 is more or less continuous at the other phase transition temperatures, excluding T_{C4}. Further, the ^1H spin-lattice relaxation time passes through a minimum value near 220 K; the presence of this minimum is attributed to the effects of rotational tunneling motion. From the ^1H T_1 curves, the relaxation process is affected by molecular motion, as described by the Bloembergen-Purcell-Pound (BPP) theory [71]. From the ^1H T_1 data shown in figure 7, the correlation time τ_H for the reorientation of NH_4 ions in $(NH_4)_2ZnCl_4$ can be calculated in phase VI. The proton spin-lattice relaxation is controlled mainly by the H-H interdipolar interaction modulated by NH_4 reorientation, yielding the following equation for T_1 [72]:

$$1/T_1 = (\mu_0/4\pi)^2 (\gamma^2\hbar/r^3)^2 \{\tau_H / [1 + (\omega_H \tau_H)^2] + 4\tau_H/[1 + (2\omega_H \tau_H)^2]\}, \tag{4}$$

where γ is the gyromagnetic ratio for the ^1H nucleus, $\hbar=h/2\pi$ (h: Planck's constant), r is the proton-proton distance, and ω_H is the proton Larmor frequency. The minimum in T_1 occurs when $\omega_H\tau_H$ =0.616. Because the T_1 curve contains a minimum, it is possible to determine the value of the constant in the BPP formula. The correlation time can be calculated as a function of temperature from that constant by using the proton Larmor frequency $\omega_H/2\pi$=200 MHz and the values of T_1 obtained from our experimental results. The correlation time for the proton nuclei is shown as a function of inverse temperature in figure 8. The correlation time τ_H of ^1H for a thermally activated reorientation follows the usual Arrhenius expression $\tau_H=\tau_o\exp(-E_a/k_BT)$, where k_B is the Boltzmann constant; therefore, the activation energy E_a and pre-exponential factor τ_o can be determined from the slopes of the straight line portions of the semilog plots of the correlation time vs. 1000/T. The activation energy in the low-temperature range was found to be 29.95 ± 0.85 kJ/mol for ^1H; this T_1 minimum at lower temperatures is related to NH_4^+ rotational tunneling motion. The activation energy of the molecular motion for the high temperature side of this minimum is 10.99 ± 0.37 kJ/mol.

The intensity of the ^1H NMR resonance lines was found to be more sensitive than the relaxation time to the phase transitions in this crystal. The ^1H T_1 was observed to vary continuously with temperature without jumps or changes. These smooth transitions demonstrate that the phase changes in

$(NH_4)_2ZnCl_4$ crystals are not of the first-order type. The high activation energy in $(NH_4)_2ZnCl_4$ at low temperatures suggests that there is a considerable degree of rotational freedom for NH_4^+ ions.

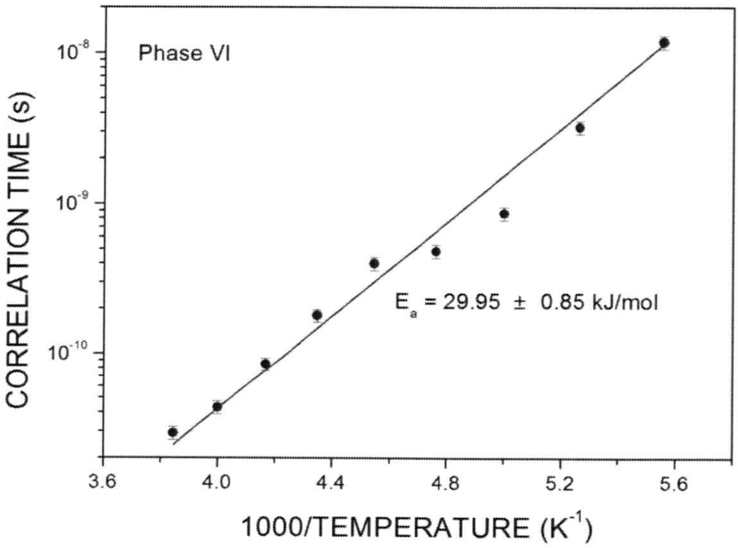

Figure 8. Correlation time for ^1H in phase VI of $(NH_4)_2ZnCl_4$ crystal as a function of inverse temperature.

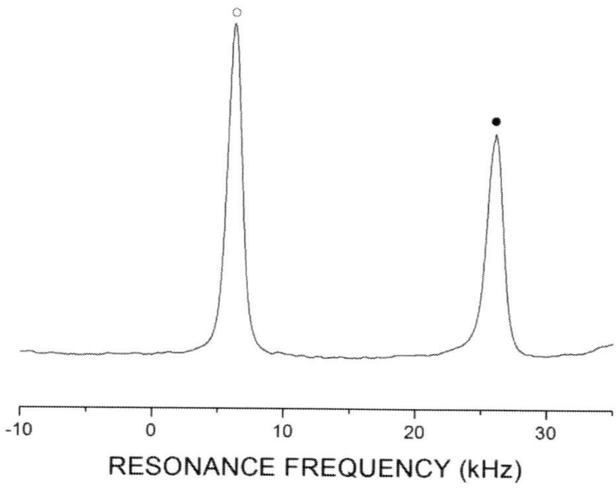

Figure 9. ^{87}Rb NMR spectrum of the crystallographically inequivalent Rb ions in Rb_2ZnCl_4 crystal [○: Rb(1) and ●: Rb(2)].

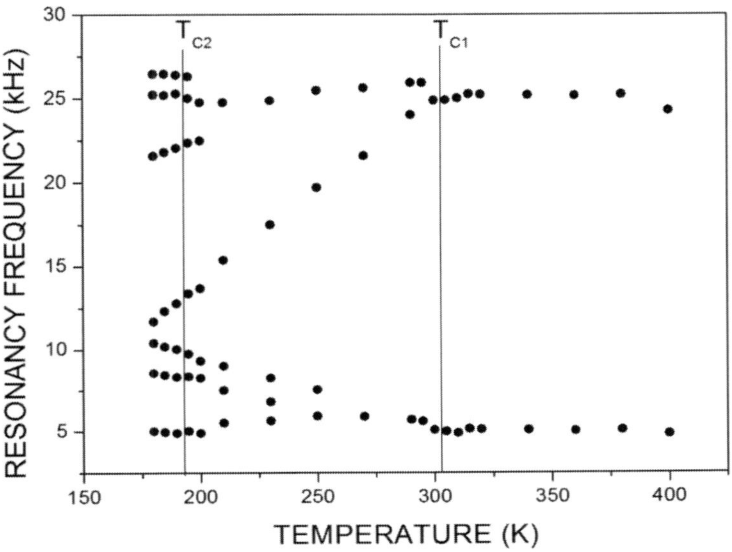

Figure 10. Splitting of Rb resonance lines as a function of temperature.

4.2. ^{87}Rb NMR Spectrum and Spin-lattice Relaxation Time in Rb_2ZnCl_4 Crystals

The NMR spectrum of ^{87}Rb (I=3/2) in Rb_2ZnCl_4 crystals was obtained at a frequency of $\omega_o/2\pi$=130.93 MHz. When these crystals are rotated about the crystallographic axis, crystallographically equivalent nuclei would be expected to give rise to three lines: one central line and two satellite lines. The NMR spectrum of ^{87}Rb (I=3/2) consists of two central lines at room temperature, as shown in figure 9. The magnitudes of the quadrupole parameters of ^{87}Rb nuclei are on the order of megahertz, so usually only central lines are obtained. The satellite resonance lines for the ^{87}Rb nucleus corresponding to transitions between levels (+2/3 $\leftrightarrow$ +1/2) and (−1/2 $\leftrightarrow$ −2/3) are out of the frequency range of the NMR probe. Thus, the two resonance lines represent the central transition of the ^{87}Rb NMR spectrum. These two Rb signals are the ^{87}Rb central NMR lines due to the two inequivalent Rb nuclei, Rb(1) and Rb(2). The spectrum consists of two lines displaced to higher frequency relative to the reference signal, which was an ^{87}Rb line obtained from an aqueous solution of $RbNO_3$. With decreasing temperatures, the lines broaden significantly and their intensity decreases. The typical temperature dependence of the second-order quadrupole shifts of the two central ^{87}Rb NMR lines is shown in figure 10. In the paraelectric phase ($T > T_{C1}$), two lines are observed. Below T_{C1}, these resonance lines split. When the crystal is cooled, further splitting of the resonance lines is observed. When the temperature decreases below 192 K, all the ^{87}Rb resonance lines are broadened, and new lines are detected in addition to the initial five lines.

The nuclear magnetization recovery curves of the ^{87}Rb nuclei were obtained by measuring the nuclear magnetization after applying saturation pulses at several temperatures. The recovery trace for the central resonance line of ^{87}Rb, with dominant quadrupole relaxation, can be represented by a combination of the two exponential functions in equation (2). The temperature dependences of the ^{87}Rb spin-lattice transition rates W_1 and W_2 were obtained from equation (2). W_1 is nearly equal to W_2 over the entire temperature range. If W_1 and W_2 have the same values in the recovery traces of ^{87}Rb in Rb_2ZnCl_4 crystals, the relaxation time T_1 can be defined according to $T_1=1/(2W_1)$. When $W_1 \neq W_2$, the constant $5/[2(W_1 + 4W_2)]$ is introduced instead of T_1; this constant is identical to T_1 when $W_1=W_2$. The variations in the relaxation times of the central lines for Rb(1) and Rb(2) with temperature were measured. The trends in the relaxation times for Rb(1) and Rb(2) are very similar; their values are consistent within the error range. The T_1 values for the ^{87}Rb nuclei as a function of inverse temperature are shown in figure 11. The ^{87}Rb relaxation times for a Rb_2ZnCl_4 crystal at the two Larmor frequencies of 65.48 and 130.93 MHz are nearly frequency independent. The relaxation times for the ^{87}Rb nuclei change significantly near 75 K ($=T_{C3}$), 192 K ($=T_{C2}$), and 303 K ($=T_{C1}$), which indicates that the spin dynamics are dramatically altered at the transition temperatures. Thus, the phase transition near T_{C3} is of the second-order type, and the phase transitions near T_{C2} and T_{C1} are of the first-order type [38]. The relaxation time decreases in phases III and IV as the temperature is increased, whereas in phase II it is nearly constant with increasing temperature. The anomalous decrease in T_1 around T_{C3} is due to the critical slowing down of the soft mode as it is usually observed in structural phase transitions from the monoclinic to the orthorhombic phase in this crystal. The jump in T_1 at T_{C2} and the behavior of T_1 throughout phase II are, however, rather unusual. The abrupt change in T_1 at T_{C2} can be considered a shortening in the a-direction as a result of a phase transition from the commensurate to the incommensurate phase; above T_{C2}, the short T_1 means that the energy transfer rate from the nuclear spin system to the surrounding environment is easier than that below T_{C2}.

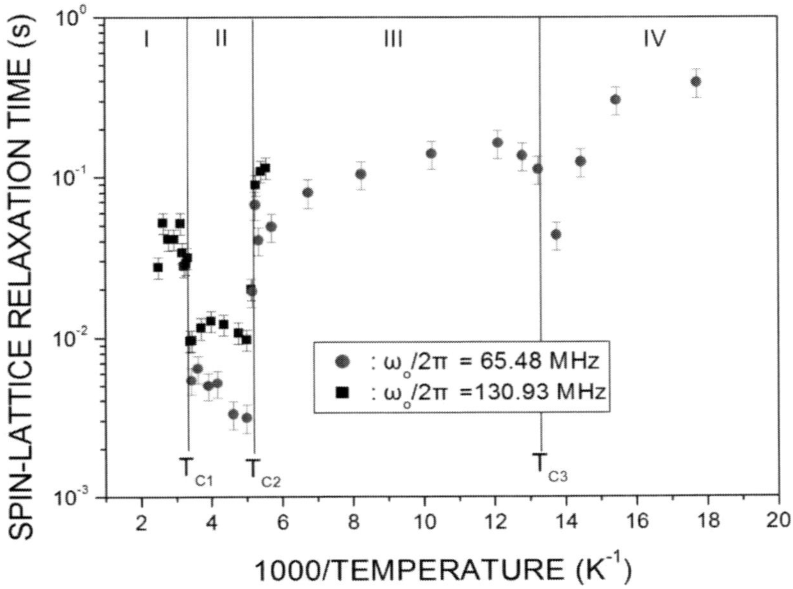

Figure 11. Spin-lattice relaxation time T_1 of ^{87}Rb nuclei in Rb_2ZnCl_4 crystal as a function of inverse temperature (●: 65.48 MHz and ■: 130.93 MHz).

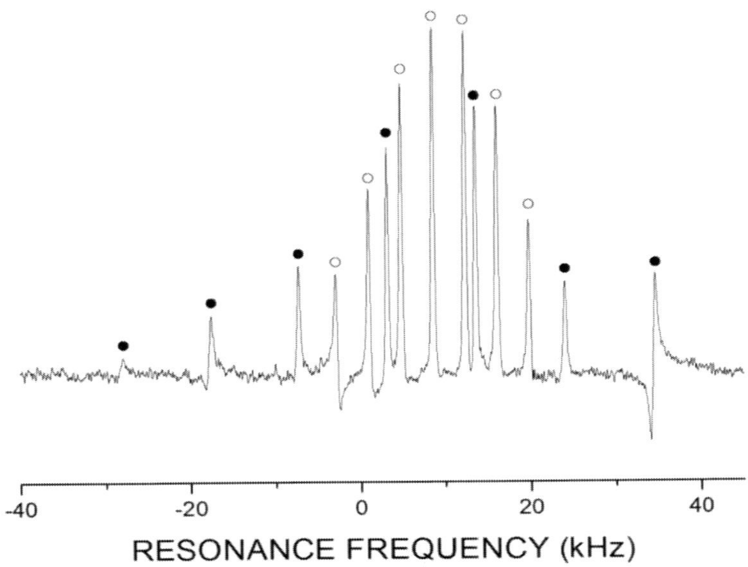

Figure 12. ^{133}Cs NMR spectrum of Cs_2ZnCl_4 crystal at room temperature. Static magnetic field B_0 is parallel to the a-axis [○: Cs(1) and ●: Cs(2)].

4.3. ^{133}Cs NMR Spectrum and Spin-lattice Relaxation Time in Cs_2ZnCl_4 Crystals

The NMR spectrum of ^{133}Cs (I=7/2) along the crystallographic a-axis at room temperature is shown in figure 12. Two groups of seven resonance lines consisting of a strong central line and six well resolved satellites were observed. The spectra of the two groups are associated with the two crystallographically inequivalent positions of the cesium atoms in the unit cell. The central line is more intense than the other lines. The zero point of the *x*-axis is the resonance frequency, 52.48 MHz, of the ^{133}Cs nucleus. The Cs spectrum is displaced by a chemical shift to higher frequencies relative to the reference signal obtained for the ^{133}Cs line from an aqueous solution of $CsNO_3$. The shifts of the central lines in these two groups, one larger and the other smaller, correspond to the transitions of the ^{133}Cs NMR lines due to Cs(1) and Cs(2), respectively. The results for Cs(1) and Cs(2) are shown as open and solid circles, respectively, in figure 12. The spacing between the resonance lines for the two groups decreases with increasing temperature, as shown in figure 13. The central transition remains almost unshifted over the temperature range 180-420 K, and the resonance line splitting between the satellite lines was found to vary with the temperature. The separation of the resonance lines for Cs(1) is small, whereas that for Cs(2) is large. The variations with temperature in the line positions indicate that the electric field gradient (EFG) at the ^{133}Cs sites varies with temperature.

The nuclear magnetization recovery traces for the central lines of Cs(1) and Cs(2) were measured at several temperatures, and it was found that the saturation recovery traces could be represented by a combination of four exponential functions, as in equation (3). The traces are shown as a function of the delay time at room temperature in figure 14. The ^{133}Cs relaxation time was obtained in terms of W_1 [$T_1=1/(1.333W_1)$], and its temperature dependence is shown in figure 15. The relaxation times of Cs(1) and Cs(2) decrease with increasing temperature. Here, the T_1 values of Cs(1) are longer than those of Cs(2). In the polyhedron of the Cs(1) ions, the average Cs–Cl distance is 3.563 Å, and in that of the Cs(2) ion, it is 3.494 Å, which is considerably smaller. This is qualitatively consistent with the longer bond length of Cs(1)–Cl and the shorter Cs(2)–Cl bond length. The ^{133}Cs nucleus in Cs_2ZnCl_4 has a very long relaxation time of approximately 600 *s* at room temperature. Further, this relaxation time does not vary abruptly with temperature, so this material does not exhibit a structural phase transition within the investigated temperature range. The NMR results clearly distinguished the Cs(1) and Cs(2) sites.

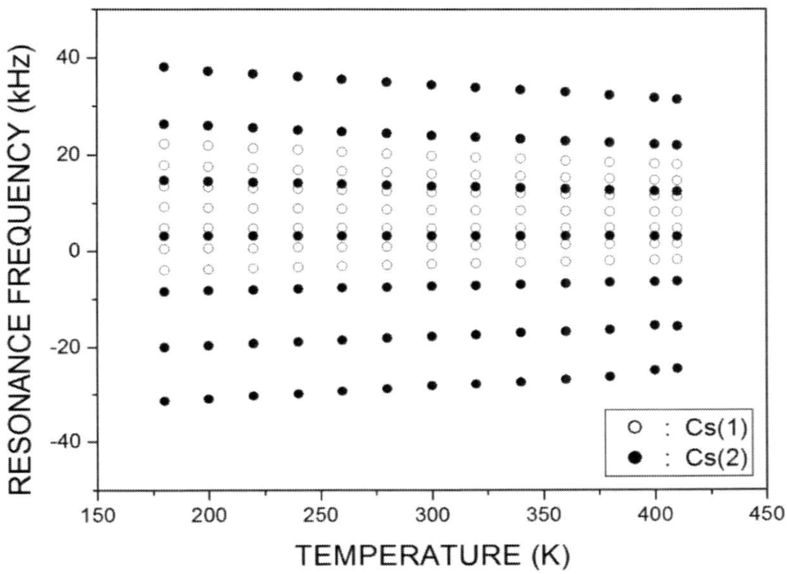

Figure 13. Splitting of Cs resonance lines as a function of temperature in Cs_2ZnCl_4 crystal.

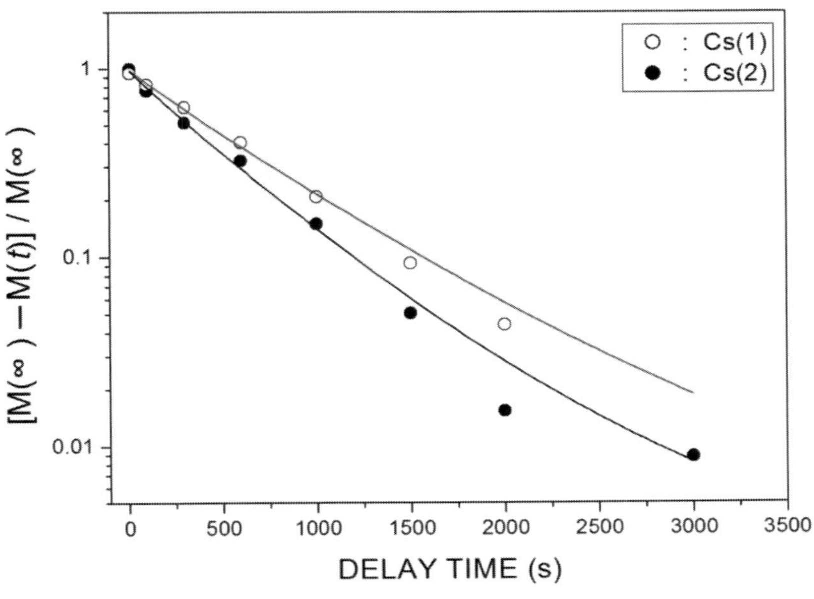

Figure 14. Recovery behaviors for Cs(1) and Cs(2) as a function of delay time at 300 K in Cs_2ZnCl_4 crystal.

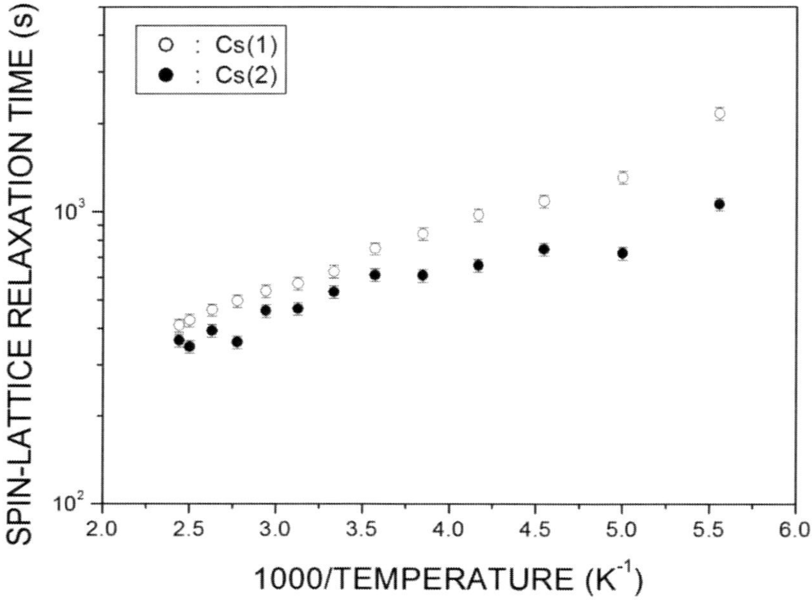

Figure 15. Spin-lattice relaxation time T_1 of ^{133}Cs nuclei in Cs_2ZnCl_4 crystal.

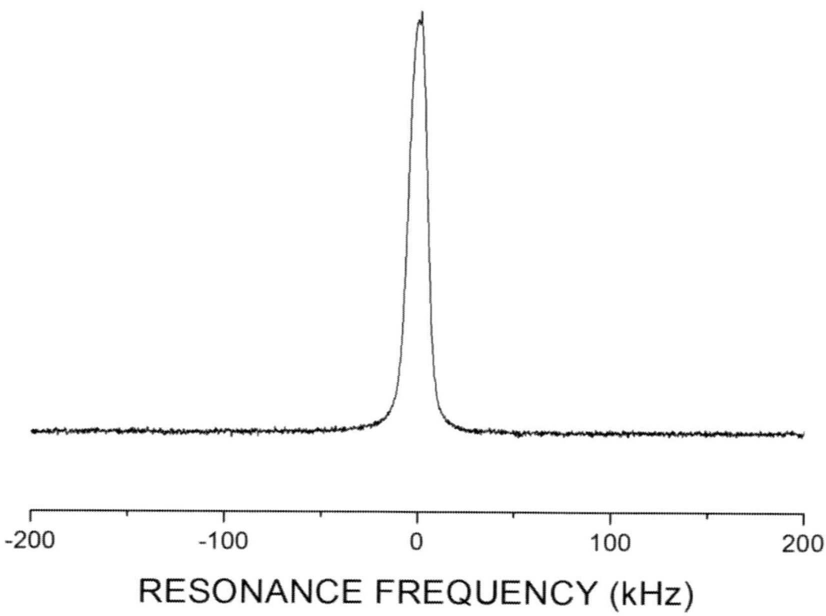

Figure 16. ^1H NMR spectrum of $NH_4RbZnCl_4$ crystal at room temperature.

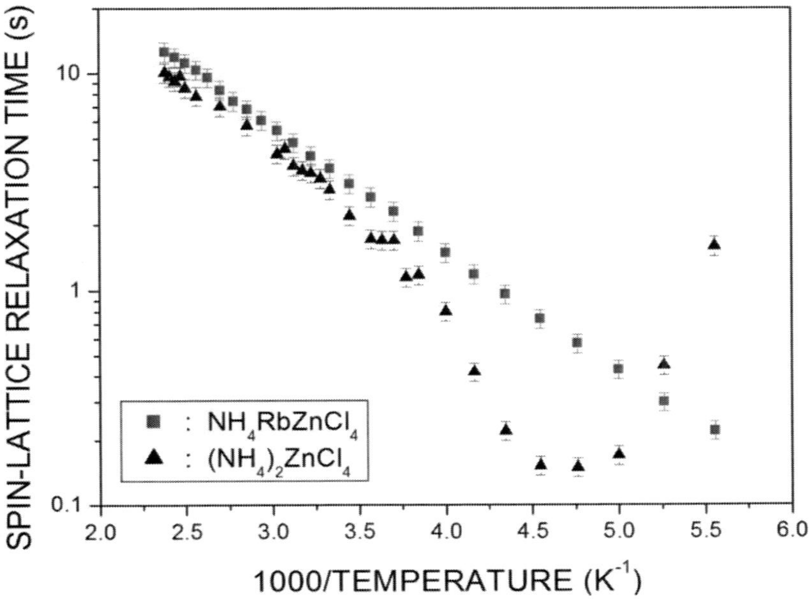

Figure 17. [1]H spin-lattice relaxation time T_1 in $NH_4RbZnCl_4$ and $(NH_4)_2ZnCl_4$ crystals as a function of inverse temperature.

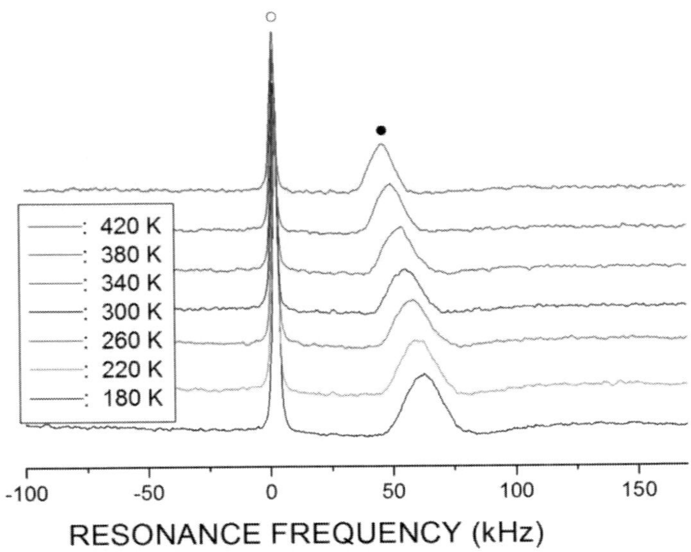

Figure 18. Splitting of [87]Rb resonance lines in $NH_4RbZnCl_4$ crystal as a function of temperature [●: Rb(1) and ○: Rb(2)].

4.4. ^{1}H and ^{87}Rb NMR in $NH_4RbZnCl_4$ Crystals

The x-ray diffractometry measurements of the mixed $NH_4RbZnCl_4$ crystals revealed an orthorhombic structure with cell parameters of a = 8.626 Å, b = 8.667 Å, and c = 14.643 Å. In the DSC experiment, no phase transitions were detected within the investigated temperature range of 170–570 K.

The ^{1}H NMR spectrum of $NH_4RbZnCl_4$ at room temperature is shown in figure 16. In $(NH_4)_2ZnCl_4$, the spectra of the two groups are associated with the two crystallographically inequivalent $NH_4(1)$ and $NH_4(2)$ positions of ammonium atoms in the unit cell, as shown in figure 4. In contrast, the ^{1}H NMR spectrum for ammonium ions in $NH_4RbZnCl_4$ mixed crystals shows only one resonance line, as shown in figure 16. The zero point of the *x*-axis corresponds to the resonance frequency of the ^{1}H nucleus, 200 MHz.

The magnetization recovery traces of the ^{1}H nuclei in $NH_4RbZnCl_4$ crystals were measured as a function of the delay times. From these results, the ^{1}H spin–lattice relaxation times T_1 of $(NH_4)_2ZnCl_4$ and $NH_4RbZnCl_4$ were obtained as a function of inverse temperature, as shown in figure 17. The ^{1}H T_1 in $NH_4RbZnCl_4$ increases with increasing temperature. Further, this relaxation time does not vary abruptly with temperature, so this mixed crystal does not exhibit a structural phase transition within the investigated temperature range. The T_1 trends of the $(NH_4)_2ZnCl_4$ and $NH_4RbZnCl_4$ crystals differ for temperatures below 250 K, but are similar for temperatures above 250 K. One possible reason for the difference is the difference between the rotational modes of the NH_4^+ ions in these crystals.

The NMR spectra of ^{87}Rb (I = 3/2) in $NH_4RbZnCl_4$ crystals were obtained at a frequency of $\omega_o/2\pi$=130.93 MHz. Two Rb central resonance lines were obtained when the magnetic field was applied along the crystallographic c-axis. The ^{87}Rb NMR spectrum of $NH_4RbZnCl_4$ is shown in figure 18 as a function of temperature. The two Rb signals are the ^{87}Rb central NMR lines due to the two inequivalent Rb nuclei, Rb(1) and Rb(2). On the other hand, the NMR spectrum of ^{87}Rb (I = 3/2) in Rb_2ZnCl_4 consists of two central lines at room temperature. The nuclear magnetization recovery curves of the ^{87}Rb nuclei in $NH_4RbZnCl_4$ were obtained by measuring the nuclear magnetization after applying saturation pulses at several temperatures. The recovery trace for the central resonance line of ^{87}Rb, with dominant quadrupole relaxation, can be represented by a combination of the two exponential functions in equation (2). The spin–lattice relaxation T_1 for ^{87}Rb was obtained in terms of W_1 and W_2, and W_1 is nearly equal to W_2 over the entire temperature range. The temperature dependence of the relaxation time for Rb in $NH_4RbZnCl_4$ crystals was obtained [73], and compared with that for Rb_2ZnCl_4 crystals, as shown in figure 19. For $NH_4RbZnCl_4$, the spin–lattice relaxation time T_1 for the Rb nucleus decreases slowly with increasing temperature. The trends in the relaxation times of Rb(1) and Rb(2) are very similar, and their values are consistent within the error range. Moreover, the ^{87}Rb relaxation times for a Rb_2ZnCl_4 crystal at the two Larmor frequencies, 65.48 and 130.93 MHz, are nearly frequency-independent.

4.5. ^{1}H and ^{133}Cs NMR in $NH_4CsZnCl_4$ Crystals

The structure of the $NH_4CsZnCl_4$ crystal is orthorhombic with cell parameters a=9.769 Å, b=7.357 Å, and c=12.944 Å. The lattice constants and phase transition temperatures for $A_{2-x}B_xZnCl_4$ crystals are shown in Table 3. In addition, the DSC results were obtained: one exothermic peak and two endothermic peaks were observed at 247, 295, and 464 K, respectively, as shown in figure 20. These temperatures are related to the phase transitions: 247 K (=T_{C3}), 295 K (=T_{C2}), and 464 K (=T_{C1}).

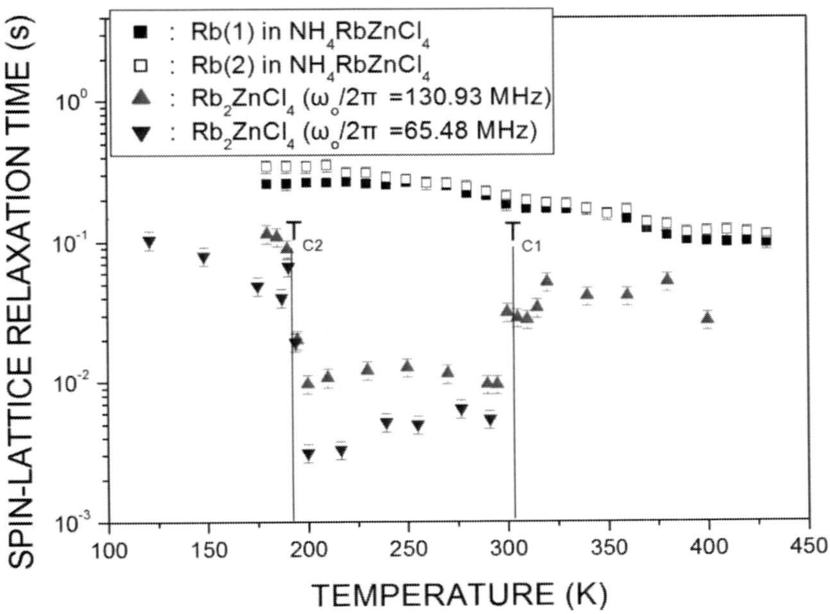

Figure 19. Temperature dependences of ^{87}Rb spin-lattice relaxation time T_1 in $NH_4RbZnCl_4$ and Rb_2ZnCl_4 crystals.

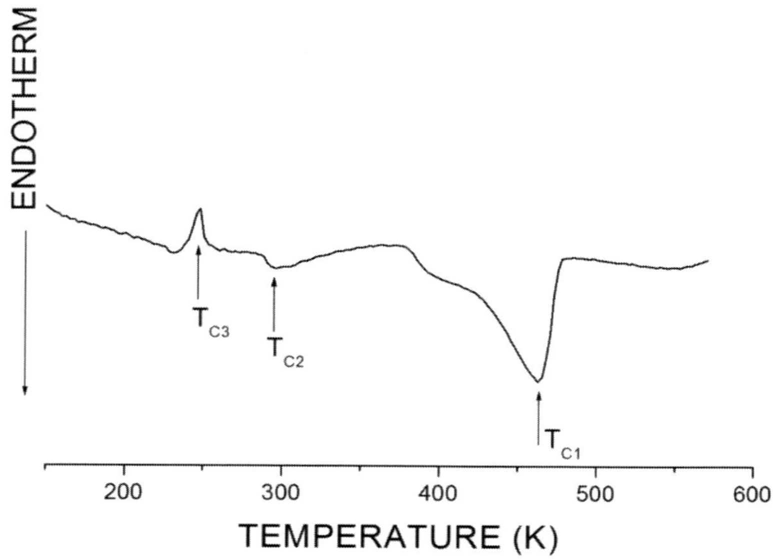

Figure 20. Differential scanning calorimetry (DSC) thermogram of $NH_4CsZnCl_4$ crystal.

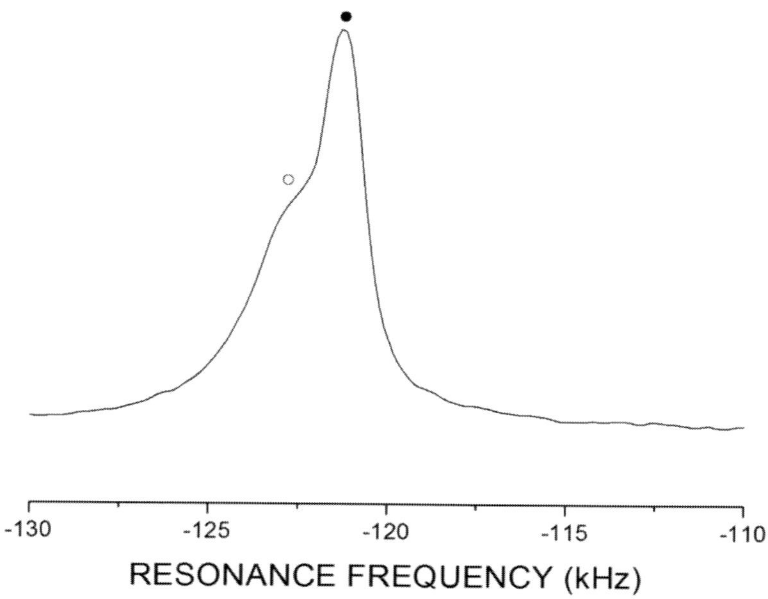

Figure 21. ^1H NMR spectrum of crystallographically inequivalent NH_4 ions in $NH_4CsZnCl_4$ crystal [●: $NH_4(1)$ and ○: $NH_4(2)$].

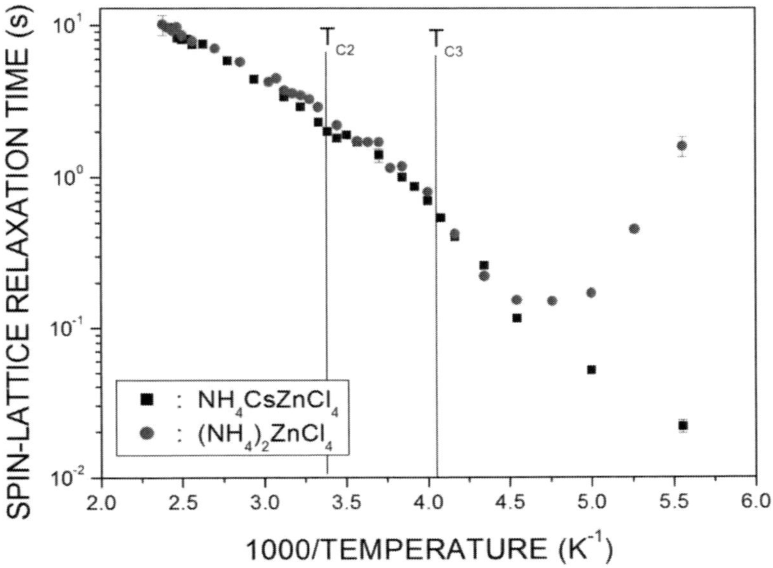

Figure 22. [1]H spin-lattice relaxation time T_1 in $NH_4CsZnCl_4$ and $(NH_4)_2ZnCl_4$ crystals as a function of inverse temperature.

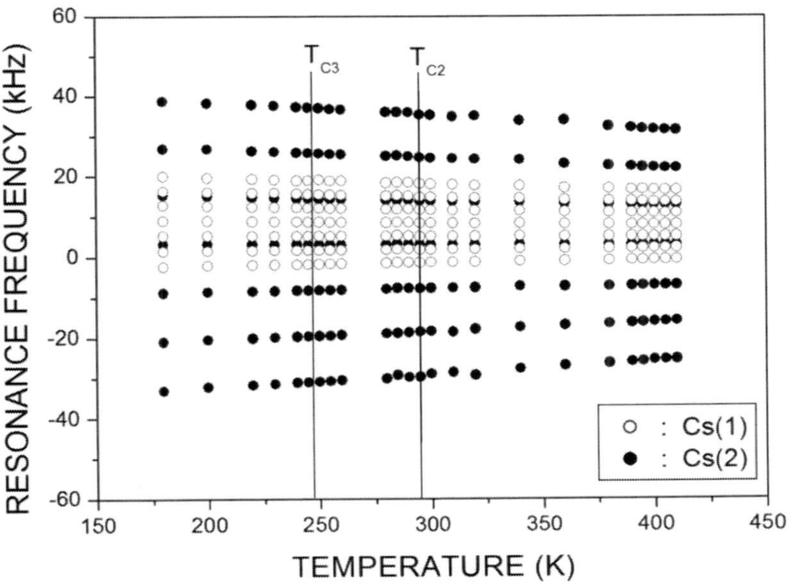

Figure 23. Splitting of Cs resonance lines as a function of temperature in $NH_4CsZnCl_4$ crystal.

Table 3. The lattice constants and the phase transition temperatures, T_C, for mixed $NH_4RbZnCl_4$, $NH_4CsZnCl_4$, and $RbCsZnCl_4$ crystals

	$NH_4RbZnCl_4$	$NH_4CsZnCl_4$	$RbCsZnCl_4$
Lattice constants	a=8.626	a=9.769	a=9.567
(Å)	b=8.667	b=7.357	b=7.347
	c=14.643	c=12.944	c=12.707
T_C (K)	None	464 K (=T_{C1})	None
		295 K (=T_{C2})	
		247 K (=T_{C3})	

The ^{1}H NMR spectrum for ammonium ions in $NH_4CsZnCl_4$ mixed crystals contains two resonance lines, as shown in figure 21. These two lines have different intensities and line widths, and are due to the two crystallographically inequivalent types of ^{1}H nuclei in the groups $NH_4(1)$ and $NH_4(2)$. The zero point corresponds to the resonance frequency, 200 MHz, of the ^{1}H nucleus.

The ^{1}H spectrum is displaced by a chemical shift to a lower frequency relative to the reference signal obtained for the ^{1}H line from an aqueous solution of H_2O. The shifts of these two lines, one smaller and the other larger, result in the ^{1}H NMR lines due to $NH_4(1)$ and $NH_4(2)$, respectively.

The ^{1}H spin-lattice relaxation time for $NH_4CsZnCl_4$ was obtained as a function of inverse temperature, as shown in figure 22. T_1 increases with increasing temperature, and the T_1 values near T_{C2} and T_{C3} are more or less continuous. Here, the T_1 ^{1}H values of $(NH_4)_2ZnCl_4$ and $NH_4CsZnCl_4$ crystals were compared. Their T_1 trends differ for temperatures below 250 K, but are similar for temperatures above 250 K.

One possible reason for the difference is the difference between the rotational modes of the NH_4^+ ions in these crystals. In $NH_4CsZnCl_4$, the activation energy below T_{C3} was determined to be 20.86 kJ/mol for ^{1}H. The activation energy of the molecular motion between T_{C3} and T_{C2} is 12.87 kJ/mol, and above T_{C2} it is 12.27 kJ/mol.

The NMR spectrum of ^{133}Cs (I=7/2) along the crystallographic a-axis in $NH_4CsZnCl_4$ is shown as a function of temperature in figure 23. The splitting of the ^{133}Cs resonance lines in $NH_4CsZnCl_4$ crystals decreases with temperature, as was the case for the Cs spectrum of Cs_2ZnCl_4. Two different groups of Cs resonance lines are evident. The spectra of the two groups are associated with the two crystallographically inequivalent positions of the cesium atoms in the unit cell. The central line is more intense than the other lines. The zero point of the y-axis is the resonance frequency, 52.48 MHz, of the ^{133}Cs nucleus.

The Cs spectrum is displaced by a chemical shift to higher frequencies relative to the reference signal obtained for the ^{133}Cs line from an aqueous solution of $CsNO_3$. The shifts of these two groups, one larger and the other smaller, result in the ^{133}Cs NMR lines due to Cs(1) and Cs(2), respectively. The results for Cs(1) and Cs(2) are shown as open and solid circles, respectively, in figure 23. Because the resonance frequency of the central line is almost constant and adjacent lines are equally spaced, the first-order perturbation of H_Q with respect to H_Z is influenced. The central transition line is not influenced by the second-order quadrupole shift. The spacing between the resonance lines for the two groups decreases with increasing temperature. The separation of the resonance lines is small for Cs(1), whereas that of the Cs(2) lines is large. The variations with temperature of the line positions indicate that the EFG at the ^{133}Cs sites varies with temperature.

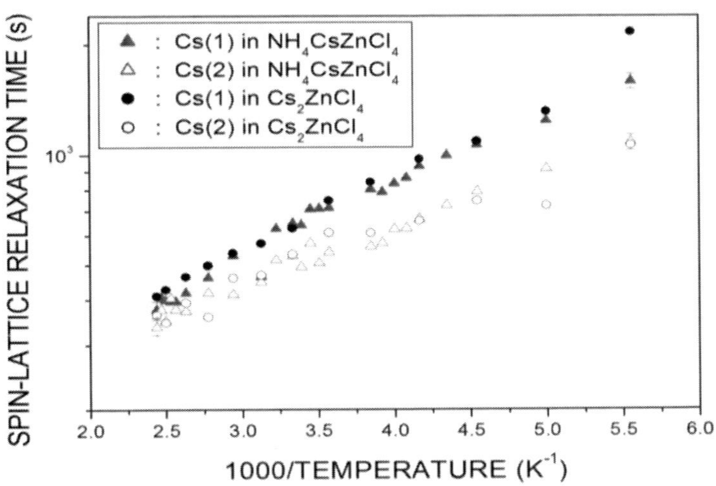

Figure 24. Spin-lattice relaxation time T_1 of crystallographically inequivalent ^{133}Cs nuclei in $NH_4CsZnCl_4$ and Cs_2ZnCl_4 crystals as a function of inverse temperature.

The recovery traces of the magnetization for the central line of ^{133}Cs can be represented by a combination of the four exponential functions in equation (3). We determined the variations with temperature of the relaxation times for the central resonance lines of the Cs(1) and Cs(2) nuclei in $NH_4CsZnCl_4$ crystals. The ^{133}Cs relaxation time, T_1, was obtained in terms of W_1 by using $T_1=1/(1.333W_1)$, and its temperature dependence is shown in figure 24.

The ^{133}Cs nucleus in $NH_4CsZnCl_4$ has a very long relaxation time of approximately 1000 *s* at room temperature [74]. This result is very similar to that for Cs in Cs_2ZnCl_4 crystals.

The T_1 values for Cs(1) are similar to those for Cs(2). Both are more or less continuous near T_{C2} and T_{C3}, and decrease with increasing temperature. These spin-lattice relaxation times are long because of the very small electric quadrupole moment; T_1 is inversely proportional to the electric quadrupole moment, Q, $[1/T_1 \propto (e^2qQ/h)^2]$ [56].

4.6. ^{87}Rb and ^{133}Cs NMR in $RbCsZnCl_4$ Crystals

The structure of $RbCsZnCl_4$ single crystals at room temperature was obtained: it is orthorhombic, with cell parameters a=9.567 Å, b=7.347 Å, and c=12.707 Å. These results are similar to those obtained for *x*=0.71 and 1.11, as reported by Kolinsky et al. [42]. In addition, no phase transitions were detected within the investigated temperature range, 170–570 K.

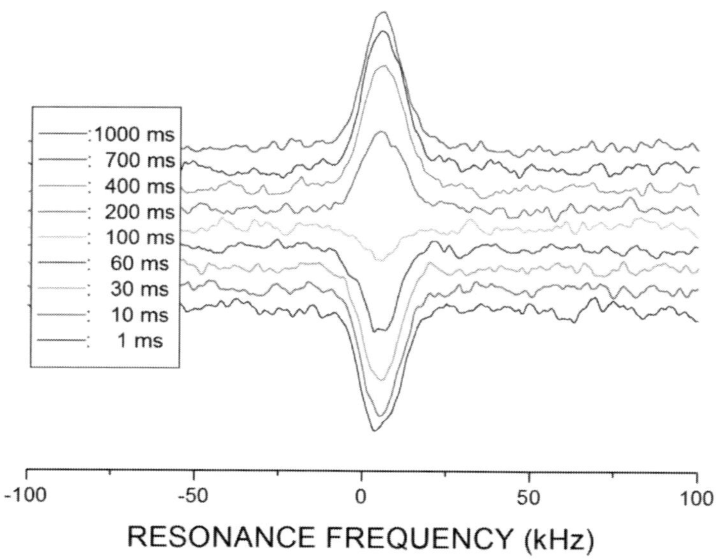

Figure 25. Inversion recovery traces for Rb as a function of delay time at 300 K in $RbCsZnCl_4$ crystal.

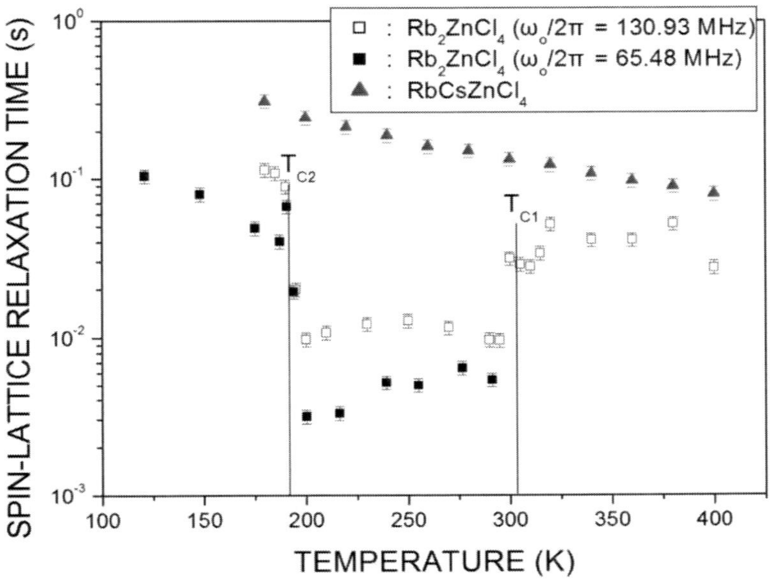

Figure 26. Temperature dependence of spin-lattice relaxation time T_1 of ^{87}Rb nuclei in $RbCsZnCl_4$ and Rb_2ZnCl_4 crystals.

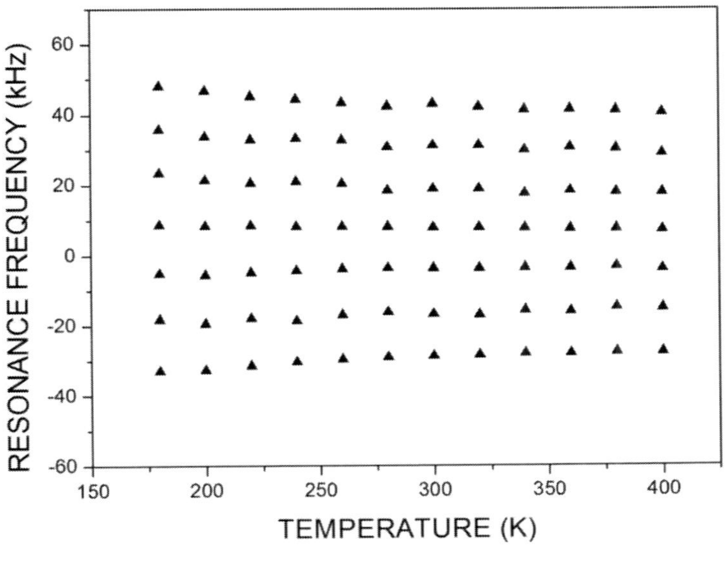

Figure 27. Splitting of Cs resonance lines as a function of temperature in $RbCsZnCl_4$ crystal.

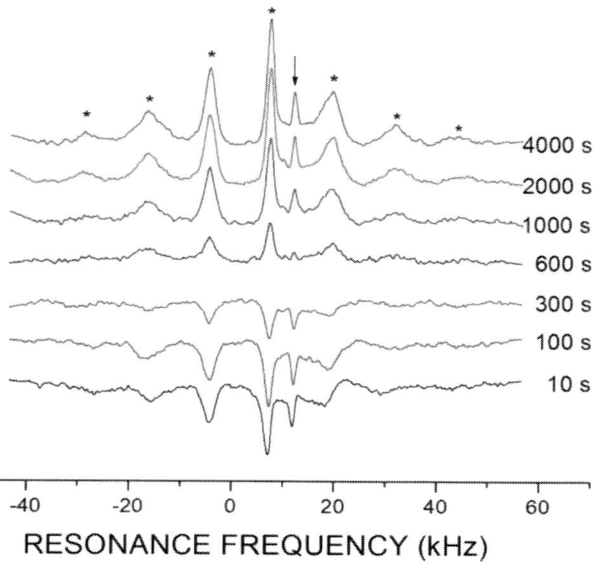

Figure 28. Inversion recovery traces for Cs as a function of delay time at 300 K in $RbCsZnCl_4$ crystal. Asterisks indicate Cs signals; arrow denotes a non-Cs signal due to Teflon rod within NMR probe.

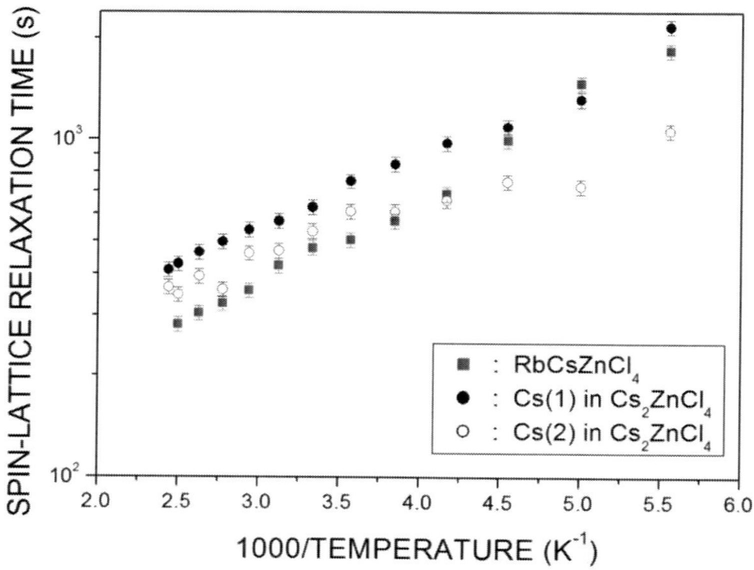

Figure 29. Spin-lattice relaxation time T_1 of ^{133}Cs nuclei in $RbCsZnCl_4$ and Cs_2ZnCl_4 crystals as a function of inverse temperature.

The NMR spectrum of ^{87}Rb (I=3/2) in $RbCsZnCl_4$ was obtained at a frequency of $\omega_o/2\pi$=130.93 MHz. When these crystals are rotated about the crystallographic axis, the crystallographically equivalent nuclei would be expected to give rise to three lines: one central line and two satellite lines. However, one Rb central resonance line was obtained for the $RbCsZnCl_4$ crystal with the magnetic field applied along the crystallographic c-axis. Nonetheless, the crystallographically inequivalent Rb sites, Rb(1) and Rb(2) were present in Rb_2ZnCl_4.

The inversion recovery traces for delay times ranging from 1 to 1000 ms at room temperature are shown in figure 25. The recovery traces for the central Rb line of $RbCsZnCl_4$ can be represented by a combination of two exponential functions, as in equation (2). The spin-lattice relaxation time T_1 for ^{87}Rb was obtained in terms of W_1 and W_2. The temperature dependences of the relaxation time of the central line for Rb in $RbCsZnCl_4$ and Rb_2ZnCl_4 single crystals are compared in figure 26. The trends of T_1 for the two crystals are very different, and T_1 for Rb in $RbCsZnCl_4$ decreases linearly with increasing temperature.

We also obtained the NMR spectrum of ^{133}Cs (I=7/2) in $RbCsZnCl_4$ single crystals at several temperatures. Seven resonance lines, consisting of a strong central line and six well-resolved satellite lines, were observed when the magnetic field was applied along an arbitrary direction. The zero point of the y–axis in figure 27 is the resonance frequency of the ^{133}Cs nucleus, 52.48 MHz. The central transition remains almost unshifted by the quadrupole interaction over the temperature range 180-400 K, and the resonance line splitting between the satellite lines was found to vary with temperature. The splitting of the ^{133}Cs resonance line decreases with temperature, as shown in figure 27. The variations with temperature in the line positions indicate that the EFG at the ^{133}Cs sites varies with temperature [75].

The recovery traces of the magnetization were determined for this crystal for delay times ranging from 10 to 4000 s, as shown in figure 28. Here, the asterisks indicate the Cs signals, and the arrow denotes a non-Cs signal due to the Teflon rod within the NMR probe. The variation with temperature of the relaxation time for the central resonance line of the ^{133}Cs nuclei was determined. The ^{133}Cs relaxation time, T_1, was obtained in terms of W_1 by using equation (3). Its temperature dependence is shown in figure 29, which also shows the T_1 values for ^{133}Cs in Cs_2ZnCl_4 crystals for comparison. T_1 for ^{133}Cs in $RbCsZnCl_4$ and Cs_2ZnCl_4 crystals has very similar values. Here, the spin-lattice relaxation time T_1 is long because of the very small electric quadrupole moment. T_1 continuously decreases with increasing temperature.

5. Discussion

We compared the *A* (=NH_4, Rb, and Cs) NMR results for A_2ZnCl_4 single crystals. The phase transition temperatures of $(NH_4)_2ZnCl_4$ are 266, 271, 319, 364, and 406 K, and those of Rb_2ZnCl_4 are 75, 192, and 303 K. Cs_2ZnCl_4 does not undergo any phase transitions. The trend in T_1 for ^1H in $(NH_4)_2ZnCl_4$ crystals differs from those for Rb and Cs in Rb_2ZnCl_4 and Cs_2ZnCl_4 crystals, respectively; rotational tunneling motion occurs in $(NH_4)_2ZnCl_4$, but not in Rb_2ZnCl_4 and Cs_2ZnCl_4. Thus, the A_2ZnCl_4 single crystals differ in their phase transition temperatures and molecular motions. The difference between the spin-lattice relaxation times of the A_2ZnCl_4 single crystals can be attributed to the different electric quadrupole moments of Rb and Cs. The molecular motions and phase transition behaviors of these crystals can also be explained in terms of the masses of their respective ions: NH_4^+ ions are lighter than Rb^+ and Cs^+ ions. Further, as the radii of the metal ions increase, a corresponding increase in the spin-lattice relaxation times of the *A* nuclei is observed. The differences observed in the spin-lattice relaxation times of the *A* nuclei when the alkali-metal ion changes may be related to the ionic radius. This result suggests that the differences in the chemical properties of *A* are responsible for the variations in the molecular motions and phase transition properties observed in the crystals studied. Therefore, it seems that the occurrence of phase transitions in these materials essentially depends on the presence of the metal ions.

The crystallographic structure of mixed $NH_4RbZnCl_4$ crystals was analyzed for the first time on the basis of knowledge of $(NH_4)_2ZnCl_4$ and Rb_2ZnCl_4 crystals. In $(NH_4)_2ZnCl_4$ crystals, crystallographically inequivalent NH_4 ions, $NH_4(1)$ and $NH_4(2)$, are present at the A_1 and A_2 sites, respectively [29]. Moreover, in Rb_2ZnCl_4 crystals, crystallographically inequivalent Rb ions, Rb(1) and Rb(2), are present at the A_1 and A_2 sites, respectively [34]. However, in $NH_4RbZnCl_4$, the distribution of NH_4 and Rb among the two cationic sites A_1 and A_2 was found to be random; according to the NMR results, crystallographically equivalent NH_4 and crystallographically inequivalent Rb ions contributed to the NMR spectrum. The Rb and NH_4 occupation rates of the two available sites, A_1 and A_2, were determined from the NMR spectrum, as shown in Table 4. The occupation rates of the two different sites in $NH_4RbZnCl_4$ mixed crystals are 1 and 0 for $NH_4(1)$ and $NH_4(2)$, respectively, and 1 and 1 for Rb(1) and Rb(2), respectively [73]. In the $NH_4RbZnCl_4$ crystal, ammonium ions only partially replace rubidium ions, causing variations in the symmetry of the crystal structure.

Table 4. *A* and *B* occupation rates of the two different sites $A_{2-x}B_xZnCl_4$ (*A* = NH_4, Rb, Cs; *B* = Rb, Cs; *x* = 0, 1, and 2) single crystals

Crystals	A_1 site		A_2 site	
	$NH_4(1)$	Rb(1)	$NH_4(2)$	Rb(2)
$(NH_4)_2ZnCl_4$	1	0	1	0
$NH_4RbZnCl_4$	1	1	0	1
Rb_2ZnCl_4	0	1	0	1
Crystals	A_1 site		A_2 site	
	$NH_4(1)$	Cs(1)	$NH_4(2)$	Cs(2)
$(NH_4)_2ZnCl_4$	1	0	1	0
$NH_4CsZnCl_4$	1	1	1	1
Cs_2ZnCl_4	0	1	0	1
Crystals	A_1 site		A_2 site	
	Rb(1)	Cs(1)	Rb(2)	Cs(2)
Rb_2ZnCl_4	1	0	1	0
$RbCsZnCl_4$	0	1	1	0
Cs_2ZnCl_4	0	1	0	1

A NMR study of $NH_4CsZnCl_4$ crystals was performed and the results were compared with those for pure $(NH_4)_2ZnCl_4$ and Cs_2ZnCl_4 crystals. The crystallographically inequivalent NH_4 ions, $NH_4(1)$ and $NH_4(2)$, in $(NH_4)_2ZnCl_4$ crystals are present at the A_1 and A_2 sites, respectively. Further, in Cs_2ZnCl_4 crystals the crystallographically inequivalent Cs ions, Cs(1) and Cs(2), are present at the A_1 and A_2 sites, respectively. However, for $NH_4CsZnCl_4$ mixed crystals, the contributions to the NMR spectrum of the crystallographically inequivalent sites $NH_4(1)$ and $NH_4(2)$, Cs(1) and Cs(2), were obtained.

The Cs and NH_4 occupation rates of the two available sites, A_1 and A_2, were determined from the NMR signals, as shown in Table 4. The distribution of NH_4 and Cs among the two cationic sites A_1 and A_2 was found to be random; according to the NMR results, the occupation rates of the two different sites of $NH_4CsZnCl_4$ mixed crystals are 1 and 1 for Cs(1) and Cs(2), respectively, and 1 and 1 for $NH_4(1)$ and $NH_4(2)$, respectively [75].

For mixed $RbCsZnCl_4$ crystals, the crystallographic structure was analyzed on the one basis of knowledge of the ^{87}Rb and ^{133}Cs NMR spectra. Crystallographically inequivalent Rb sites, Rb(1) and Rb(2), are present in Rb_2ZnCl_4 crystals, and crystallographically inequivalent Cs sites, Cs(1) and Cs(2), are present in Cs_2ZnCl_4 crystals. However, only one Rb site and only one Cs site are present in $RbCsZnCl_4$ crystals, according to previously reported x-ray results. Our NMR experimental results confirmed these conclusions, as shown in Table 4. These results are consistent with the crystal structures of Kolinsky et al. [42]. The Cs and Rb occupation rates of the two available sites, A_1 and A_2, were determined from the NMR signals [74]; the occupation rates of the two different sites in $RbCsZnCl_4$ crystals are 0 and 1 for Rb(1) and Rb(2), respectively, and 1 and 0 for Cs(1) and Cs(2), respectively.

The NMR spectra and spin-lattice relaxation times, T_1, of their ^1H, ^{87}Rb, and ^{133}Cs nuclei were determined using NMR spectroscopy. According to our experimental results, the spin-lattice relaxation times, T_1, of Cs-based compounds are very much longer than those of Rb-based compounds. The differences between the T_1 values of the two nuclei can be attributed to their different Larmor and quadrupole frequencies, among other factors. This difference is due to the difference between the electric quadrupole moments of Rb and Cs (Q_{Rb}: 13.35×10^{-30} m^2 and Q_{Cs}: -0.343×10^{-30} m^2); the spin-lattice relaxation time is inversely proportional to the square of the quadrupole coupling constant. The magnitude of the term e^2qQ/h for the ^{87}Rb nuclei is on the order of megahertz, whereas that of the ^{133}Cs nuclei is on the order of kilohertz. In simple NMR theory, the typical behavior of the spin-lattice relaxation rate for random motions of the Arrhenius type with a correlation time τ_c is described in terms of three regimes, including both, fast, and slow motion regimes. The fast motion regime can be described as $\omega_o\tau_c \ll 1$, $T_1^{-1} \sim \exp[E_a/RT]$, and the slow motion regime can be described as $\omega_o\tau_c \gg 1$, $T_1^{-1} \sim \omega_o^{-2}\exp[-E_a/RT]$, where ω_o is the Larmor frequency, and E_a is the activation energy. It is known that different limits are satisfied for $\omega_o\tau_c$ in pure A_2ZnCl_4 and mixed $A_{2-x}B_xZnCl_4$ crystals. Specifically, the limit $\omega_o\tau_c \gg 1$ would apply for ^{87}Rb in $NH_4RbZnCl_4$ and $RbCsZnCl_4$ crystals, and for ^{133}Cs in Cs_2ZnCl_4, $NH_4CsZnCl_4$, and $RbCsZnCl_4$ crystals. Further, the limit $\omega_o\tau_c \ll 1$ would apply for ^1H in $NH_4RbZnCl_4$ and $NH_4CsZnCl_4$ crystals.

CONCLUSION

The pure crystals A_2ZnCl_4 and mixed crystals $A_{2-x}B_xZnCl_4$ (A = NH_4, Rb, Cs; B = Rb, Cs; x = 0, 1, and 2) were grown using the slow evaporation method, and their crystal structures were determined using x-ray diffraction. To obtain information about the occupation probabilities governing the partial replacement of *A* ions with *B* ions, an NMR study of $A_{2-x}B_xZnCl_4$ was performed, and the results were compared with those for pure A_2ZnCl_4 crystals.

The lattice constants of $A_{2-x}B_xZnCl_4$ differ greatly from those of pure A_2ZnCl_4 crystals, and the phase transition temperatures also differ from those of pure crystals. Although the ionic radii of *B* ions is similar to that of *A* ions (Rb ion: 1.47 Å, NH_4 ion: 1.43 Å, and Cs ion: 1.67 Å), the addition of a small amount of *B* ions changes the transition temperature. In the mixed $A_{2-x}B_xZnCl_4$ crystals, the *A* ions only partially replace the *B* ions, and consequently cause variations in the symmetry of the crystal structure.

The crystallographic structure of mixed crystals $A_{2-x}B_xZnCl_4$ was compared with that of pure A_2ZnCl_4 crystals. In the pure crystals, there are two crystallographically inequivalent NH_4 sites, $NH_4(1)$ and $NH_4(2)$, in $(NH_4)_2ZnCl_4$; two crystallographically inequivalent Rb sites, Rb(1) and Rb(2), in Rb_2ZnCl_4; and two crystallographically inequivalent Cs sites, Cs(1) and Cs(2), in Cs_2ZnCl_4.

However, the occupation rates of the two different sites in $NH_4RbZnCl_4$ mixed crystals are 1 and 0 for $NH_4(1)$ and $NH_4(2)$, respectively, and 1 and 1 for Rb(1) and Rb(2), respectively. For $NH_4CsZnCl_4$, the distribution of NH_4 and Cs among the two cationic sites A_1 and A_2 was found to be random; according to the NMR results, the occupation rates of the two different sites of $NH_4CsZnCl_4$ mixed crystals are 1 and 1 for Cs(1) and Cs(2), respectively, and 1 and 1 for $NH_4(1)$ and $NH_4(2)$, respectively. Further, only one Rb site and only one Cs site appear in $RbCsZnCl_4$. The Cs and Rb occupation rates of each of the two available sites, A_1 and A_2, were determined from the NMR signals.

An NMR study of mixed $A_{2-x}B_xZnCl_4$ crystals was performed to obtain information about the occupation probabilities governing the partial replacement of *A* ions with *B* ions. Consequently, the structural properties of $A_{2-x}B_xZnCl_4$ crystals have been investigated, and the results regarding the occupation rates of the *A* and *B* nuclei were analyzed.

ACKNOWLEDGMENT

This research was supported by the Basic Science Research program through the National Research Foundation of Korea (NRF) funded by the Ministry of Education, Science, and Technology (2012001763).

REFERENCES

[1] Tylczynski, Z.; Piskunowicz, P. *Phys. Stat. Solidi* (a) 1990, 118, 441-446.
[2] Gillet, A. M.; Luspin, Y.; Hauret, G. *Solid State Commun*. 1987, 64, 797-799.
[3] Cummins, H. Z. *Phys. Reports* 1990, 185, 211-409.
[4] Koyama, Y.; Nagata, T.; Koike, K. *Phys. Rev*. B 1995, 51, 12157-12164.
[5] Matsunaga, H. *J. Phys. Soc. Japan* 1982, 51, 864-872.
[6] Matsunaga, H. *J. Phys. Soc. Japan* 1982, 51, 873-879.
[7] Burris, J. L.; Tereshchenko, L.; Kabos, P.; Hochheimer, H. D.; Luspin, Y.; Evenson, C. R.; Dorhout, P. K. *J. Phys. Chem. Solids* 2000, 61, 719-726.
[8] Tylczynski, Z.; Karaev, A. D. *Physica* B, 2001, 307, 224-230.
[9] Misul, S. V.; Belobrova, I. A. *Phys. Stat. Solidi* (a) 1982, 70, K167-170.
[10] Amin, M.; El-Korashy, A.; Ibrahim, S. S.; Dabour, M.; Arafa, S. *J. Phys. Chem. Solids* 2002, 63, 869-874.
[11] Sata, T.; Osaka, T.; Makita, Y. *J. Phys. Soc. Japan* 1984, 53, 1907-1910.
[12] Matsunaga, H.; Nakamura, E. *J. Phys. Soc. Japan* 1981, 50, 2789-2790.
[13] van Konigsveld, H. *Acta Cryst*. C 1983, 39, 15-19.
[14] Sawada, S.; Shiroishi, Y.; Yamamoto, A.; Takashige, M.; Matsuo, M. *J. Phys. Soc. Japan* 1977, 43, 2099-2010.
[15] Gesi, K.; Iizumi, M. *J. Phys. Soc. Japan* 1979, 46, 697-698.
[16] Francke, E.; Postollec, M. Le.; Mathieu, J. P.; Poulet, H. *Solid State Commun*. 1980, 33, 155-156.
[17] Noiret, I.; Guinet, Y.; Hedoux, A. *Phys. Rev*. B 1995, 52, 13206-13213.
[18] Wada, M.; Sawada, A.; Ishibashi, Y. *J. Phys. Soc. Japan* 1979, 47, 1185-1190.
[19] Wada, M.; Sawada, A.; Ishibashi, Y. *J. Phys. Soc. Japan* 1981, 50, 531-537.
[20] Lu, H. M.; Hardy, J. R. *Phys. Rev*. B 1992, 45, 7609-7620.

[21] Lamba, O. P.; Sinha, S. K. *Solid State Commun.* 1986, 57, 365-371.
[22] Wong, P. T. T. *J. Chem. Phys.* 1976, 64, 2186-2191.
[23] Sundaram, C. S.; Ramakrishna, J.; Chandrasekhar, K.; Sastry, V. S. S. *Ferroelectrics* 1986, 69, 299-311.
[24] Vuorimaki, A. H.; Punkkinen, M.; Ylinen, E. E. *Chem. Phys. Lett.* 1989, 161, 561-566.
[25] Ingman, L. P.; Punkkinen, M.; Vuorimaki, A. H.; Ylinen, E. E. *J. Phys. C: Solid State Phys.* 1985, 18, 5033-5041.
[26] Michel, D.; Muller, B.; Petersson, J.; Trampert, A.; Walisch, R. *Phys. Rev.* B 1991, 43, 7507-7518.
[27] Liu, Z. S.; Punkkinen, M.; Vuorimaki, A. H.; Ylinen, E. E. *Physica* B 1994, 202, 343-345.
[28] Stoch, G.; Ylinen, E. E.; Punkkinen, M.; Petelenz, B.; Birczynski, A. *Solid State Nucl. Mag. Reson.* 2009, 35, 180-186.
[29] Lim, A. R. *Solid State Commun.* 2011, 151, 1631-1634.
[30] Schneider, E. *Solid State Commun.* 1982, 44, 885-887.
[31] Blinc, R.; Aleksandrova, I. P.; Chaves, A. S.; Milia, F.; Rutar, V.; Seliger, J.; Topic, B.; Zumer, S. *J. Phys. C: Solid State Phys.* 1982, 15, 547-563.
[32] Blinc, R.; Milia, F.; Rutar, V.; Zumer, S. *Phys. Rev. Lett.* 1982, 48, 47-50.
[33] Rutar, V.; Milia, F.; Topic, B.; Blinc, R.; Rasing, Th. *Phys. Rev.* B 1982, 25, 281-289.
[34] Blinc, R.; Prelovsek, P.; Kind, R. *Phys. Rev.* B 1983, 27, 5404-5411.
[35] Petersson, J.; Schneider, E. *Z. Phys. B-Condens. Matter* 1985, 61, 33-44.
[36] Ailion, D. C.; Chen, S. *Solid State Commun.* 1994, 92, 835-837.
[37] Walisch, R.; Petersson, J.; Schubler, D.; Hacker, U.; Michel, D.; Perez-Mato, J. M. *Phys. Rev.* B 1994, 50, 16192-16204.
[38] Lim, A. R.; Jung, J. K.; Jeong, S. Y. *Solid State Commun.* 2001, 118, 453-457.
[39] Subramanian, R. K.; Venu, K.; Sastry, V. S. S. *J. Phys.: Condens. Matter* 1994, 6, 137-146.
[40] Weakliem, H. A. *J. Chem. Phys.* 1962, 36, 2117-2140.
[41] Lim, A. R.; Han, O. H.; Jeong, S. Y. *J. Phys. Chem. Solids* 2003, 64, 933-937.
[42] Kolinsky, C.; Jannin, M.; Puget, R.; Delarue, P.; Godefroy, G. *J. Phys.: Condens. Matter* 1997, 9, 825-843.
[43] SMART and SAINT-Plus v 6.22, Bruker AXS Inc., Madison, Wisconsin, USA, 2000.

[44] Belobrova, I. A.; Moskalev, A. K.; Bizukina, N. V.; Milul, S. V.; Aleksandrova, I. P. *Solid State Commun.* 1980, 33, 1101-1102.
[45] Warczewski, J.; Broda, H. *Phase Transitions* 1981, 2, 131-140.
[46] Matsunaga, H.; Itoh, K.; Nakamura, E. *Acta Cryst.* B 1982, 38, 898-900.
[47] de Bruin-Hordijk, G. J.; Peterse, W. J. A. M.; Laman, C. D.; Berg, A. J. V. D.; Tuinstra, F. *Ferroelectrics* 1993, 140, 281-286.
[48] Andrews, S. R.; Mashiyama, H. *J. Phys. C: Solid State Phys.* 1983, 16, 4985-4996.
[49] Parisi, F.; Bonadeo, H. *Acta Cryst.* A 1997, 53, 286-290.
[50] Hedoux, A.; Grebille, D.; Jaud, J.; Godefroy, G. *Acta Cryst.* B 1989, 45, 370-378.
[51] Secco, A. S.; Trotter, J. *Acta Cryst.* 1983, C39, 317-318.
[52] Edwardson, P. J.; Katkanant, V.; Hardy, J. R.; Boyer, L. L. *Phys. Rev.* B. 1987, 35, 8470-8477.
[53] Edwardson, P. J.; Hardy, J. R. *Phys. Rev.* B 1988, 38, 2250-2256.
[54] McGinnety, J. A. *Inorg. Chem.* 1974, 13, 1057-1061.
[55] Puget, R.; Jannin, M.; Perret, R.; Godefroy, L.; Godefroy, G. *Ferroelectrics* 1990, 107, 229-234.
[56] Abragam, A. *The Principles of Nuclear Magnetism*; Oxford University Press, Oxford, 1961.
[57] Cowan, B. *Nuclear Magnetic Resonance and Relaxation*; Cambridge University Press, Cambridge, 1997.
[58] Slichter, C. P. *Principles of Magnetic Resonance*; Springer, Berlin, 1989.
[59] Lim, A. R.; Ichikawa, M. *J. Phys.: Condensed Matter* 2006, 18, 2173-2180.
[60] Lim, A. R. *J. Phys.: Condensed Matter* 2007, 19, 116216-116228.
[61] Lim, A. R.; Shin, H. *J. Appl. Phys.* 2010, 107, 63513-63518.
[62] Fukushima, E.; Roeder, S. B. W. *Experimental Pulse NMR*; Addition-Wesley, Reading, MA, 1981.
[63] van der Klink, J. J.; Rytz, D.; Borsa, F.; Hochli, U. T. *Phys. Rev.* B 1983, 27, 89-101.
[64] Harris, R. K. *Nuclear Magnetic Resonance Spectroscopy*; Pitman Pub. Inc., London, 1983.
[65] Lim, A. R.; Ichikawa, M. *J. Solid State Chem.* 2006, 179, 117-121.
[66] Lim, A. R. *Phys. Rev.* B 2005, 72, 64103-64108.
[67] Tewari, D. P.; Verma, G.S. *Phys. Rev.* 1963, 129, 1975-1979.
[68] Gordon, M. I.; Hoch, M. J. R. *J. Phys.* C 1978, 11, 783-795.

[69] Kim, K. H.; Torgeson, D. R.; Borsa, F.; Martin, S. W. *Solid State Ionics* 1996, 90, 29-38.
[70] Lim, A. R. *Solid State Nucl. Magn. Reson*. 2009, 36, 45-51.
[71] Bloembergen, N.; Purcell, E. M.; Pound, R. V. *Phys. Rev*. 1948, 73, 679-712.
[72] Bohmer, R.; Jeffrey, K. R.; Vogel, M. *Prog. Nucl. Magn. Reson. Spectrosc*. 2007, 50, 87-174.
[73] Lim, A. R.; Lim, K. Y. *J. Solid State Chem*. 2013, 200, 227-231.
[74] Lim, A. R. *J. Appl. Phys*. 2010, 107, 93508-93513.
[75] Lim, A. R. *Chem. Phys*. 2012, 400, 39-43.

In: Nuclear Magnetic Resonance (NMR) ISBN: 978-1-63117-768-2
Editor: Doddapuneni Krishna Rao © 2014 Nova Science Publishers, Inc.

Chapter 9

^{1}H NMR, A TOOL FOR QUALITATIVE AND QUANTITATIVE STUDY OF FOOD LIPID COMPOSITION AND OF ITS EVOLUTION IN DEGRADATION PROCESSES

Andrea Martínez-Yusta, Encarnación Goicoechea and Maria D. Guillén*

Food Technology, Lascaray Research Center, Faculty of Pharmacy, University of the Basque Country (UPV/EHU), Vitoria, Spain

ABSTRACT

This chapter focusses on the usefulness of Proton Nuclear Magnetic Resonance (^{1}H NMR) spectroscopy in the study of edible oils and fats, and of food lipids in general, not only from a qualitative but also from a quantitative point of view.

As far as compositional studies are concerned, it has been evidenced by several authors that this technique provides a great deal of detailed information concerning the main components of food lipids, which is to say triglycerides, as well as about other minor compounds such as phospholipids, cholesterol, sterols, etc. Moreover, several approaches have been proposed to quantify these compounds by means of ^{1}H NMR spectral data.

---

* E-mail: mariadolores.guillen@ehu.es.

This technique has also been employed to study the evolution of food lipid degradation processes, providing information not only about their primary oxidation compounds, such as hydroperoxides and related conjugated dienic systems, but also about their secondary oxidation compounds, such as aldehydes. It is noteworthy that ^{1}H NMR makes it possible to detect and quantify toxic oxygenated *alpha, beta*-unsaturated aldehydes, such as 4-hydroxy-2-alkenals, generated by the degradation of polyunsaturated omega-3 and omega-6 acyl groups, and considered potential causal agents of several diseases, including cancer, Alzheimer and Parkinson. Studies carried out by ^{1}H NMR have shown that, depending on both the processing conditions and food lipid composition, great differences can be found regarding both the rate of their degradation and the concentration and nature of the compounds generated.

Keywords: ^{1}H NMR, food lipids, oils and fats, triglycerides, minor food lipids, characterization, quantification, oxidation, thermo-degradation, primary oxidation compounds, secondary oxidation compounds, oxygenated *alpha, beta*-unsaturated aldehydes

1. Introduction

Lipids form an important group of nutrients that are found in numerous foods and are the main constituents of some of them, such as edible oils and fats. They play important role in food processing, at both an industrial and a domestic scale, as well as in nutrition, with noticeable effects on health. Their performance in processing and human metabolism depends on their composition, for which reason their characterization, as well as information about their oxidation level and other attributes such as oxidative stability, is essential.

Different kinds of compounds are included within the category of food lipids, the majority of them triglycerides. In addition, other compounds such as free fatty acids, mono- and di-glycerides, phospholipids, sterols, tocopherols and certain hydrocarbons like squalene, which are present in foods in small concentrations, are also food lipids. Compounds in this second group usually appear in foods together with compounds from the first group.

Classically, triglycerides have been characterized by determining the proportions of the different kinds of acyl chains by gas chromatography (GC), by means of their methyl ester derivatives which are obtained after transesterification with methanol. The oxidation level of the main food lipids, like oils and fats, has been monitored by parameters such as Peroxide Value,

Conjugated Dienes, absorption at 233 and 273 nm in ultraviolet (UV) spectroscopy, p-Anisidine value, or Tiobarbituric acid (TBA) test. These same parameters have also been used to monitor the evolution of the above mentioned lipids in oxidation or thermo-degradation processes.

The unsaponifiable lipids like sterols, tocopherols, squalene and other isoprene derivatives have usually been characterized in foods by gas or liquid chromatography, after previous extraction. The evolution in the oxidation processes of many of these compounds is usually monitored by the decrease of their concentration in the sample, although in some cases the formation of oxy-derivatives has also been studied, usually using the above cited chromatographic techniques.

Although some of these methodologies are still being used, spectroscopic techniques, especially Proton Nuclear Magnetic Resonance (^{1}H NMR), are currently gaining relevance in this field. This technique provides information about the different kinds of protons present in food lipids, giving a complete view of them in a brief period of time, and without chemical modification of the sample [1]. This information encompasses the composition in both major and minor components of food lipids, such as vegetable oils of different nature [2-4], butter and margarine [5, 6] or fish lipids [7-13].

Another subject of great interest is the influence of both food processing conditions and food lipids composition on the rate and mechanisms of their oxidation or thermo-degradation processes, as well as on the nature and concentrations of the intermediate and final compounds formed [7, 14-26]. This subject is of great importance because these processes not only generate compounds with undesirable sensorial properties, but also toxic ones that affect the safety of the foods; among the broad range of compounds which may be formed in these processes, aldehydes are considered some of the most important. In recent years special attention has been paid to oxygenated *alpha, beta*-unsaturated aldehydes, which are well-known genotoxic and cytotoxic compounds [27, 28]. In addition, it has recently been established that degradative processes of food lipids can occur not only during processing, but also during digestion [29]. ^{1}H NMR spectroscopy has proved to be also very useful in the study of all these topics.

This chapter focusses on the usefulness of the ^{1}H NMR technique to deal with the above mentioned aspects of food lipids. In a previous paper [30] the state of the art of the application of ^{1}H NMR spectroscopy to study edible oils and fats was reviewed. In recent years advances in this field have been published and all of them are summarized in this chapter.

2. Qualitative and Quantitative Characterization of Triglycerides

As commented on before, triglycerides are the main components of most food lipids, the latter containing complex mixtures of different types of triglycerides. The differences among them are due to the length, degree and type of unsaturation of their acyl groups. Figure 1 shows an example of a hypothetical triglyceride having linoleic (diunsaturated ω-6), oleic (monounsaturated ω-9) and linolenic (triunsaturated ω-3) acyl group chains. The mixtures of triglycerides have classically been characterized in function of the proportions of their several acyl groups, determined as methyl ester derivatives obtained after transesterification.

The ^{1}H NMR spectra of food triglycerides show a reduced number of signals because the environment of many of their protons is similar in the different triglycerides and acyl group chains. Figure 2 shows the ^{1}H NMR spectra of three edible oils (virgin linseed VL, sunflower SF and extra virgin olive EVO), each with a very different composition in triglycerides, and thus in acyl groups (see in Table 1 the molar percentages of the different kinds of acyl groups of these oils). As this figure shows, only eleven different signals can be distinguished; in addition to them, the spectra of the fish lipids triglycerides show additional signals due to specific acyl groups, such as docosahexaenoic (DHA), eicosapentaenoic (EPA) and arachidonic (ARA).

The correct assignment of spectral signals is essential in order to carry out an accurate characterization of the triglycerides contained in each lipid sample. Table 2 gives the chemical shifts, multiplicities and assignment of the most common ^{1}H NMR signals of the protons present in triglycerides of food lipids from vegetable and animal sources obtained from 400 MHz equipment. As some signals are specific to certain acyl groups, their identification, as well as an approximate idea of the relative proportions of some acyl groups in relation to others, is directly achievable from the spectrum.

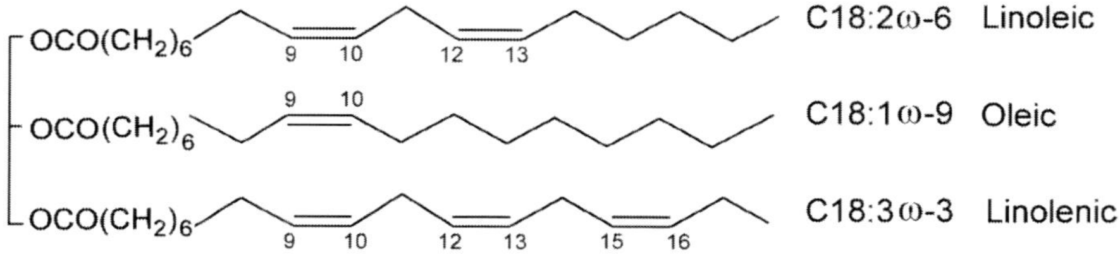

Figure 1. Structure of a possible triglyceride supporting different acyl groups.

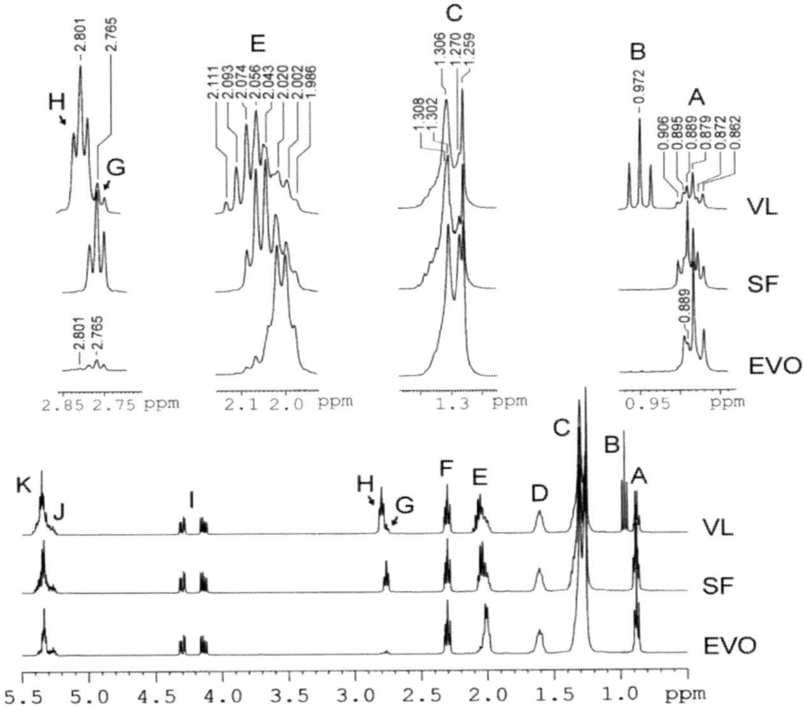

Figure 2. ^{1}H NMR spectra of three non-degraded vegetable oils: virgin linseed (VL), sunflower (SF) and extra virgin olive (EVO), together with some regions properly enlarged.

Table 1. Molar percentages of the main acyl groups of the edible oils shown in Figure 2, obtained from ^{1}H NMR data. Data taken from [24-26]

Oils	Ln (%)	L (%)	O (%)	S (%)
Virgin linseed (VL)	50.37 ± 0.50	17.85 ± 0.03	22.56 ± 0.37	9.20 ± 0.19
Sunflower (SF)	---	55.05 ± 4.31	34.80 ± 2.83	10.03 ± 0.95
Extra virgin olive (EVO)	0.49 ± 0.02	4.49 ± 0.28	82.06 ± 0.98	12.93 ± 0.65

Acyl groups: Ln, linolenic; L, linoleic; O, oleic; S, saturated.

Table 2. Chemical shifts and multiplicities of most common ^{1}H NMR signals of the spectra of triglycerides of food lipids

Chemical shift (ppm)	Multiplicity	Functional group
0.88	t	$-C\underline{H}_3$ (saturated, monounsaturated ω-9 and/or ω-7 acyl groups)
0.89	t	$-C\underline{H}_3$ (unsaturated ω-6 acyl groups)
0.97	t	$-C\underline{H}_3$ (unsaturated ω-3 acyl groups)
1.22-1.42	m*	$-(C\underline{H}_2)_n-$ (acyl groups)
1.52-1.70	m	$-OCO-CH_2-C\underline{H}_2-$ (acyl groups except for DHA, EPA and ARA acyl groups)
1.66-1.74	m	$-OCO-CH_2-C\underline{H}_2-$ (EPA and ARA acyl groups)
1.94-2.14	m**	$-C\underline{H}_2-CH=CH-$ (acyl groups except for $-CH_2-$ of DHA acyl group in β-position in relation to carbonyl group)
2.31	dt	$-OCO-C\underline{H}_2-$ (acyl groups except for DHA acyl groups)
2.36-2.42	m	$-OCO-C\underline{H}_2-C\underline{H}_2-$ (DHA acyl groups)
2.77	t	$=HC-C\underline{H}_2-CH=$ (diunsaturated ω-6 acyl groups)
2.80	t	$=HC-C\underline{H}_2-CH=$ (triunsaturated ω-3 acyl groups)
2.77-2.90	m	$=HC-C\underline{H}_2-CH=$ (DHA, EPA, ARA and other polyunsaturated ω-3 acyl groups)
4.14, 4.30	dd,dd	$-C\underline{H}_2OCOR$ (glyceryl groups)
5.20-5.26	m	$>C\underline{H}OCOR$ (glyceryl groups)
5.26-5.47	m	$-C\underline{H}=C\underline{H}-$ (acyl groups)

t: triplet, m: multiplet; *overlapping of multiplets of methylenic protons in the different acyl groups either in β-position,□ or further, in relation to double bonds, or in γ-position,□ or further, in relation to the carbonyl group; DHA: docosahexaenoic acyl groups; EPA: eicosapentaenoic acyl groups; ARA: arachidonic acyl groups; **overlapping of multiplets of the α-methylenic protons in relation to a single double bond of the different unsaturated acyl groups; d: doublet.

The signal between 0.83 and 0.93 ppm, named signal A in Figure 2, is formed by the overlapping of the triplet of the methylic protons of saturated and monounsaturated ω-9 acyl groups (plus monounsaturated ω-7 groups in certain fish lipids) centred at 0.879 and the triplet of methylic protons of unsaturated ω-6 acyl groups centred at 0.889 ppm. The fact that the area of the ^{1}H NMR signals is proportional to the number of protons that generates them (the proportions being constant for all kind of protons) allows one to extract certain quantitative information from the simple observation of the spectra. Thus, in spite of the above mentioned overlapping, simple observation of this broad signal, conveniently enlarged, shows if the molar percentage of ω-6 acyl groups predominates over that of saturated plus monounsaturated ω-9, or not. Therefore, in Figure 2 it is evident that the height of the triplet centred at 0.879

is higher than that centred at 0.889 in extra virgin olive and virgin linseed oils, and the opposite is true in sunflower oil, being in agreement with the composition in acyl groups given in Table 1.

In addition, the methylic protons of polyunsaturated ω-3 acyl groups give a triplet signal centred at 0.972 ppm (signal B in Figure 2). Simple visual comparison of the height of this signal B in the spectrum with that of the above mentioned signal A of the rest of methylic protons, allows one to extract information about the proportion of ω-3 acyl groups in the sample. This signal is very intense in the spectrum of virgin linseed oil, because this contains a very high proportion of linolenic acyl groups, in contrast with that of extra virgin olive oil, which shows a very small triplet centred at 0.972 ppm, and that of sunflower oil, which shows no signal due to the absence of linolenic acyl groups, in agreement with data in Table 1. Much more accurate information can be obtained using the area of both signals; so, the molar percentage of ω-3 acyl groups can be determined from the quotient between the area of these methylic protons (B) and the total area (A+B) [3, 8, 31, 32].

Another interesting signal is signal C in Figure 2, due to methylenic protons in the different acyl groups either in β-position, or further, in relation to double bonds, or in γ-position, or further, in relation to the carbonyl group. This signal also permits one to distinguish between saturated, monounsaturated ω-9 (oleic), and polyunsaturated (linoleic and linolenic) acyl groups, because each of these groups generates a different signal with certain overlapping. The following make a contribution to this complex signal: the methylenic protons of saturated acyl groups giving a multiplet centred at 1.259 ppm; the methylenic protons of monounsaturated ω-9 (oleic) groups which generate two multiplets, one near 1.270 ppm and another near 1.303 ppm (the latter being more intense); and the methylenic protons of diunsaturated ω-6 (linoleic) and triunsaturated ω-3 (linolenic) groups, each one giving a multiplet that overlaps with that of oleic near 1.306 ppm [2]. The height of each one of these overlapped signals can give an approximate idea of the relative abundance of monounsaturated ω-9 (oleic) *versus* saturated and polyunsaturated acyl groups in food triglycerides [2]. This is very useful for comparative purposes, so, as can be observed in Figure 2, of the three oils extra virgin olive contains the highest proportion of oleic acyl groups, due to the height of the peak at 1.270 ppm.

The methylenic protons in β-position in relation to the carbonyl group in the acyl chains of the triglycerides generate the signal D in Figure 2, centred at 1.609 ppm. The area of this signal could be useful for quantitative purposes. To this signal contributes any kind of acyl group with the two methylenic protons in β-position in relation to the carbonyl group, except for docosahexaenoic, eicosapentaenoic and arachidonic groups. These latter groups are present in fish lipids and their methylenic protons in β-position give specific signals, making them, for this reason, very useful for identification and quantification. The chemical shifts of all these signals are given in Table 2.

The *mono*-allylic protons, that is to say, the α-methylenic protons in relation to a single double bond of all unsaturated acyl groups, contribute to the generation of the broad signal E, between 1.94-2.14 ppm, formed by the overlapping of the multiplets of the different kinds of unsaturated acyl groups. Thus, the contribution of oleic acyl groups is shown by the peaks at 2.002 and at 2.020 ppm, that of linoleic by the peaks at 2.043 and 2.056 ppm and that of linolenic by the peaks at 2.074, 2.093 and 2.111 ppm [2]. As a result, simple observation of this signal can give an approximate idea of the abundance of the different kinds of unsaturated acyl groups in the food lipids subject of study. This can be corroborated by data in Table 1 and signal E in the spectra of the different oils shown in Figure 2. In addition, the area of this signal is useful for quantitative purposes to determine the molar percentage of the saturated acyl groups, by difference, because these kinds of protons are absent in them. With regard to this, it should be taken into account that any unsaturated acyl group contributes to the area of this signal E with four protons, except docosahexaenoic groups that only contribute with two (see data in Table 2).

The spectra of all triglycerides contain the signal F shown in Figure 2. This signal centred at 2.305 ppm, is due to methylenic protons in α-position in relation to the carbonyl group. All the acyl groups contribute with two protons to the area of this signal, except docosahexaenoic groups, as Table 2 indicates. This signal F does not show significant differences either in multiplicity or chemical shifts in the different kinds of acyl groups. For this reason, it does not provide information about the abundance of each kind of acyl groups separately. However, it is very useful for quantitative purposes and provides quantitative information about the total acyl groups present in the sample, except for docosahexaenoic groups, as mentioned.

Polyunsaturated acyl chains contain *bis*-allylic protons, that is to say methylenic protons in α-position in relation to two double bonds, and these generate signals between 2.72 and 2.90 ppm. Their multiplicity and chemical shifts are not the same for all polyunsaturated groups. Thus, signal G in Figure 2, due to diunsaturated ω-6 acyl groups (linoleic), is a triplet centred at 2.765 ppm, and signal H due to triunsaturated ω-3 acyl groups (linolenic) is a triplet centred at 2.801. When these two kinds of acyl groups are present in the same sample both signals overlap partially, as can be observed in Figure 2, especially in virgin linseed oil sample. The observation of this spectral region gives direct information, especially in edible oils, about their relative abundance in an approximate way; thus, Figure 2 shows that: the linoleic acyl groups are the only polyunsaturated groups in sunflower oil, in agreement with data in Table 1; the linolenic acyl groups are the main polyunsaturated acyl group in virgin linseed oil; and that they are also present in extra virgin olive oil in much smaller proportion than linoleic groups. Furthermore, this signal is very important for quantitative purposes to determine the molar percentages of linoleic and of linolenic acyl groups.

As Table 2 shows, in fish lipids samples in addition to the above mentioned polyunsaturated acyl groups, the *bis*-allylic protons of docosahexaenoic, eicosapentaenoic, arachidonic and of other polyunsaturated groups generate a broad signal in this region between 2.77 ppm and 2.90 ppm [8, 13]. This signal provides global quantitative information about all these unsaturated acyl groups, which is useful for comparative purposes. This broad signal can overlap with the triplet of linolenic and partially with the triplet of linoleic ones.

The protons of the carbons 1 and 3 of glycerol backbone generate signal I in Figure 2, (two double doublet) centred at 4.14, 4.30 ppm, whereas protons on carbon 2 generate the multiplet J, between at 5.20-5.26 ppm. The multiplicity and chemical shifts of these signals are the same in all kinds of triglycerides and for this reason they do not provide useful information to discriminate between triglycerides. Nevertheless, the area of signal I is very valuable for quantitative purposes because it is present in all triglycerides and all of them contribute to it with the same number of protons.

As Figure 2 shows, the last signal in the triglyceride spectra is signal K, which is a broad signal between 5.26-5.47 ppm generated by the olefinic protons of all the unsaturated acyl groups. This signal overlaps with the above named signal J. Each unsaturated acyl group contributes to the area of this signal K with the corresponding olefinic protons and it can be very useful for quantitave purposes. In fact, the area of this signal is a measure of the unsaturation degree of triglycerides, and so of food lipids. A very close relationship has been found between the area of this signal and the Iodine Value in oils and fats, in such a way that the equation that relates both parameters is being widely used to determine the Iodine Value, because it is a very easy way to measure this parameter [3, 23-26, 33].

Bearing in mind all the above, several approaches have been developed to determine the molar percentage of the different kinds of acyl groups in edible oils and fats, and in food lipids in general [3, 6-9, 13-26, 31, 34]. Each one of them uses the area of certain signals, and the level of accuracy, which is not very different in the various approaches, depends on the signals involved in the approach. Previous papers have mentioned the need to be careful in the establishment of the signal integration limits, in order to avoid the inclusion of side bands [2, 3]. Likewise, for quantitative purposes a relaxation delay of 2.5 s has been considered as being appropriate by some authors and it has been proved that recovery delays longer than this have little effect on the quantitative results [30, 33]; furthermore, this has been corroborated by studies carried out in our laboratory. This may be due to the fact that relaxation time can be drastically shortened by the presence of paramagnetic species in the sample, or by other causes. This fact is very important because it allows a considerable reduction in the acquisition time of the spectra.

One approach, of broad application, to determine the molar percentage of different kinds of acyl groups in vegetable oils and fats, which have the signals shown in Figure 2, involves the use of the following equations:

Molar percentage of linolenic groups, $Ln = 100\ (A_H/3A_I)$ (eq. 1)

Molar percentage of linoleic groups, $L = 100\ (2A_G/3A_I)$ (eq. 2)

Molar percentage of oleic groups, O (or MU) = $100\ [(A_E - 2A_G - A_H)/3A_I]$ (eq. 3)

Molar percentage of saturated, $S = 100\ [1 - (A_E/3A_I)]$ (eq. 4)

In these equations A_H and A_G are the areas of the signals of *bis*-allylic protons of linolenic and linoleic groups respectively, A_E is the area of *mono*-allylic protons, and A_I is the area of the protons of the glyceryl groups indicated in Figure 2 and Table 2. As has been mentioned before, other approaches have been used in the study of vegetable oils [3, 14-19, 22] and also of fish lipids [7, 8, 13], the latter being more complicated than the former due to the higher complexity of their lipids.

In summary, the ^{1}H NMR spectra of food lipids contain a great deal of information chiefly concerning their main components, that is to say their triglycerides. This information allows one to know not only the nature, but also the molar percentages of the different kinds of acyl groups, making their qualitative and quantitative characterization possible. Qualitative characterization can be achieved by simple visual observation of the spectrum, once the assignment of their signals is known. Moreover, simple visual observation of the spectrum can lead to an approximate idea of the abundances of the different kinds of acyl groups in the sample. This quantitative characterization can be made in an accurate and simple way by using the equations provided by different recently developed approaches

3. Qualitative and Quantitative Characterization of Minor Lipid Components

The use of the ^{1}H MNR spectroscopy for the identification and quantification of minor components previously extracted in a selective way from the lipid matrix is a common practice.

A much more interesting strategy is the simultaneous study of main and minor components of food lipids from the same spectrum. As ^{1}H NMR spectroscopy encompasses all the protons present in the sample, signals generated by protons of minor lipid components can be observable at the same time as triglyceride signals, as long as they do not overlap and their concentration is high enough to be detected in the used equipment. Among all the ^{1}H NMR signals of each minor food lipid component there is usually only one, or a small number of them, that meets the above mentioned conditions and can therefore be used for identificative and quantitative purposes. Table 3 shows the chemical shifts of some ^{1}H NMR signals of minor lipid components that have practical usefulness to identify and quantify them directly from food lipid spectrum. Among them there are certain acyl groups, free fatty acids, mono- and di-glycerides, phospholipids, sterols and terpenes.

The quantification of these minor components can be carried out using different strategies: to obtain absolute values of concentration the introduction of a standard compound into the sample is required; to obtain concentration values referring to main components or functional groups of the sample, such as triglycerides or total number of acyl groups, the introduction of standards is not required.

Table 3. Chemical shift assignments and multiplicities of the ^{1}H NMR signals of some minor compounds present in food lipids

Chemical shift (ppm)	Multiplicity	Functional group and compound	Reference
Minor acyl groups			
4.95-5.07	dq,dq	-CH=C$\underline{H}_2$ (ω-1 acyl groups)	[35]
5.75-5.86	m	-C$\underline{H}$=CH_2 (ω-1 acyl groups)	[35]
Fatty acids			
2.2-2.4	t	-OCO-C$\underline{H}_2$- (free fatty acids)	[36]
Mono- and Di-glycerides			
3.73	t*	-C$\underline{H}_2$OH (1,2-diglycerides)	[6]
3.66	ddd	-C$\underline{H}_2$OH (1-monoglycerides)	[6, 37]
3.95	m	-CH(OH)- (1-monoglycerides)	[6, 37]
3.83	d	-C$\underline{H}_2$OH (2-monoglycerides)	[6, 37]
4.04-4.21	m	-C$\underline{H}$(OH)- (1,3-diglycerides)	[6, 38]
Phospholipids			
3.16	s	-C$\underline{H}_2$-N (phosphatidylethanolamine)	[10]
3.32	s	-N(C$\underline{H}_3$)$_3$ (phosphatidylcholine)	[11]
Sterols			
0.540	s	-C$\underline{H}_3$ C-18 (Δ7-avenasterol)	[39]
0.678	s	-C$\underline{H}_3$ C-18 (cholesterol)	[40]
0.679	s	-C$\underline{H}_3$ C-18 (β-sitosterol and Δ5-campesterol)	[40]
0.704	s	-C$\underline{H}_3$ C-18 (Δ5-stigmasterol and brassicasterol)	[39]
Terpenes			
1.670	s	-C$\underline{H}_3$ (squalene)	[41]

d: doublet; q: quartet; m: multiplet; t: triplet; *Standard compounds (dioleoylglycerol and dipalmitin) give a triplet centred at 3.73 ppm, and in complex mixtures the signal generated is a doublet centred at 3.72 ppm instead of a triplet [6]; s: singlet.

4. Monitoring of the Degradation of Food Lipids. Identification and Quantification of Primary and Secondary Oxidation Compounds

As the technique takes in all kinds of protons present in the sample subject of study, when this undergoes chemical changes, these should be reflected in their ^{1}H NMR spectra. In the same way that the unmodified food lipids can be characterized by this technique, it should also be useful to establish the degradation level of food lipids, to monitor their evolution when submitted to oxidative or thermo-degradative conditions, and for this reason, to determine their oxidative stability.

Pioneer ^{1}H NMR studies have focussed on the oxidation of pure standard compounds, such as triolein [42], trilinolenin [43], trilinolein [44], linoleic and linolenic acids [45], methyl oleate [46], methyl linoleate [47] and its conjugated hydroperoxides coming from autoxidation [48], and methyl linolenate [49, 50] and its conjugated hydroperoxides coming from autoxidation [51], in most of the cases after extraction by preparative high performance liquid chromatography (HPLC).

Grootveld and coworkers carried out the first studies by ^{1}H NMR on the oxidation of food lipids, and not on pure standard compounds. They submitted oil samples to 180 ºC for periods of up to 90 min and also analyzed several repeatedly-used culinary frying oils [52-55].

Later on Guillén and coworkers studied, by means of ^{1}H NMR, the evolution of vegetable oils submitted to different conditions, such as storage in closed receptacles at room temperature for up to 10 years [20, 21], accelerated storage conditions in *Petri* dishes in a convection oven at 70 ºC with aeration [14-17], thermo-oxidation in *Petri* dishes in a convection oven at 100 ºC with aeration [22] and at 190 ºC with aeration [19], thermo-oxidation in *Petri* dishes in a microwave oven (at 190 ºC) [18], deep-frying conditions without food in a fryer at 190 ºC for up to 40 h [23-26] and deep-frying with different foods in a fryer at 190 ºC [34]. Furthermore, the evolution of the oxidation process of the lipids of farmed salmon fillets submitted to 50 ºC with aeration in a convection oven has also been studied by ^{1}H NMR [7].

All the above mentioned studies have shown that both the original food lipid composition and the conditions under which they are submitted decisively influence not only the rate of the degradation process, but also the nature and concentration of the compounds formed. These conclusions have been reached after a thorough study of different thermo-degradative processes of edible oils and fats, and food lipids in general.

During oil thermo-oxidation, very complex reactions take place as a result of heating and the incorporation of oxygen into the triglyceride structures. This causes the modification of the fatty acyl group chains and the generation of volatile compounds.

4.1. Monitoring of the Molar Percentage Evolution of the Different Kinds of Acyl Groups Throughout the Thermo-degradation of Food Lipids

The starting point in the study of any food lipid thermo-oxidative process is the original composition of their lipids. During degradative processes, acyl group chains are seriously affected, and for this reason the monitoring of their concentration is essential to analyze the rate of the degradative process and to know how the process affects each of the kinds of acyl groups. Carrying out this task requires the development of approaches which must be as accurate as possible to give quantitative data about the molar percentages of the acyl groups throughout the degradation process. The previous approach for non oxidized oils is also valid for oxidized oils with some provisos: equation number 4, given to determine saturated acyl groups, determines both saturated as well as modified acyl groups in the case of oxidized oils; in addition, in order to be more accurate in the case of oxidized oils, equation number 3 given to determine oleic acyl groups corresponds to monounsaturated acyl groups in general.

The determination of the evolution of the several kinds of acyl groups in the different samples has shown that there are processes, such as that provoked at 70 ºC with aeration (10 g of oil keep in *Petri* dishes), where there is an induction period before the beginning of the degradation of its acyl groups which is different for each oil. This induction period is very long for extra virgin olive oil (nearly 600 h), much shorter for sunflower oil (nearly 96 h) and almost non-existent for virgin linseed oil [15-17]. Furthermore, in this process degradation largely affects the most unsaturated groups, in such a way that they can disappear totally from the sample.

However, in other processes, such as the heating of oils under frying conditions in the absence of food (4 L of oil in an industrial fryer at 190 ºC), the degradation of acyl groups begins during the first minutes of heating; in other words there is no induction period [23-26]. The variations of the molar percentages of the different kinds of acyl groups in this process follow linear paths from the beginning, and the slopes of these lines give the rate of decrease or increase of the molar percentage of the corresponding acyl group. It has been proved that, in extra virgin olive oil, the molar percentage of the oleic group shows the highest rate of diminution; in sunflower oil this highest rate of decrease is produced in the molar percentage of the linoleic group and in virgin linseed this occurs in the molar percentage of the linolenic acyl group. The variations in the molar percentages of the other kinds of acyl groups in these oils depend on their unsaturation degree; so, in those having a higher (lower) unsaturation degree than the main acyl group a decrease (increase or no modification) is observed [25]. Furthermore, although the molar percentage of some polyunsaturated acyl groups is very small in the original oil, they do not suffer total degradation in the heating time studied.

In this kind of studies ^{1}H NMR is a much more valuable tool than the classical methodologies.

4.2. Formation and Monitoring of the So-Called Primary Oxidation Compounds

It has been postulated that the degradation of acyl groups of food lipids begins with the formation of the so called primary oxidation compounds, which include hydroperoxy groups that can be supported in structures having conjugated dienic systems. If this occurs in any food lipid thermo-oxidation process, ^{1}H NMR spectroscopy should be able to detect these functional groups, whenever their concentration is high enough to be detected by this technique.

Pioneer studies using ^{1}H NMR following preparative HPLC of the compounds formed in the oxidation of pure compounds such as triolein, trilinolein and trilinolenin at temperatures under 60 °C [43-45, 56-58], as well as studies using exclusively ^{1}H NMR concerning thermal stressing of oils and fats [52, 53] have confirmed the formation of compounds containing both the above mentioned functional groups. Recent studies of the thermo-oxidation of edible oils at 70 °C with aeration (10 g of oil in *Petri* dishes) have confirmed some of these previous results [14-17].

Figure 3 gives the spectral region in which the signals of hydroperoxy protons as well as of conjugated dienic systems associated to them, of virgin linseed, sunflower and extra virgin olive oils submitted to the above mentioned conditions for different periods of time (70 °C with aeration) appear. These spectra were recorded on a Varian 300 Plus spectrometer operating at 300 MHz [15-17]. The results obtained confirm the existence of different induction periods for each oil, as mentioned before. After this period, the formation of both kinds of functional groups, hydroperoxy (signals between 8 and 9 ppm) and *(Z,E)* and *(E,E)* conjugated dienic systems (signals between approximately 5.4 and 6.6 ppm) is observed; the detailed chemical shifts and multiplicities of these signals are given in Table 4.

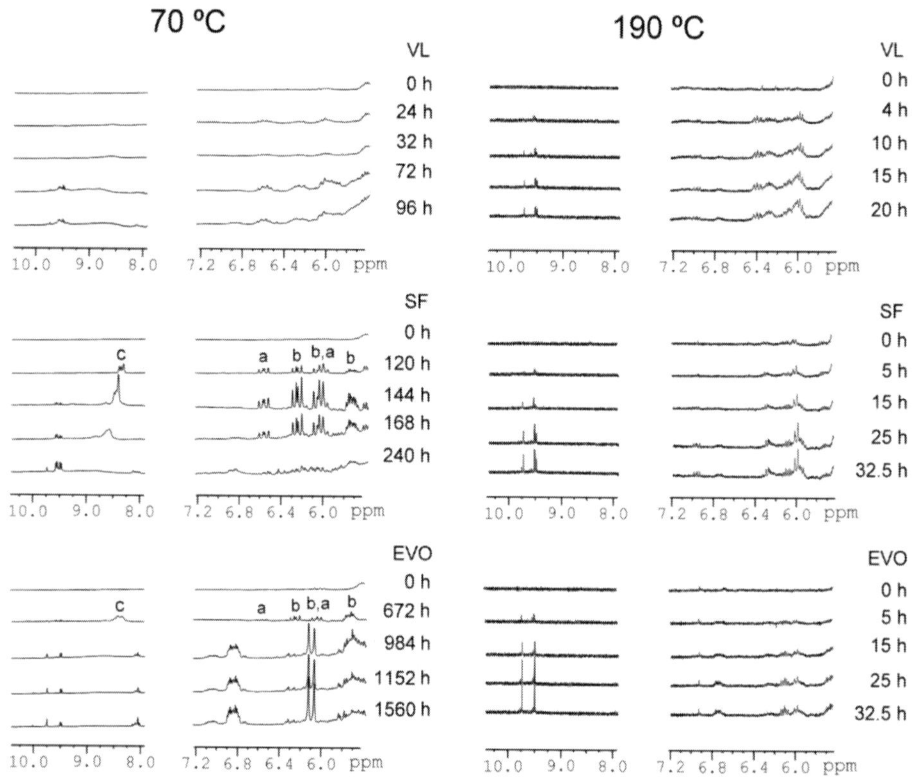

Figure 3. ^{1}H NMR spectra of VL, SF and EVO oils submitted to two thermo-oxidation processes: at 70 ºC in a convection oven and at 190 ºC in a deep-fryer. Region between 5.6-7.2 ppm, in which protons of dienic conjugated systems appear. Region between 8.0-10.0 ppm, in which hydroperoxides and aldehydic protons signals appear. The signal letters agree with those in Table 4.

Table 4. Chemical shifts and multiplicities of some of the ^{1}H NMR signals of protons of hydroperoxy groups, conjugated dienic systems, and aldehydes generated in virgin linseed, sunflower and extra virgin olive oils thermo-oxidation processes, either at 70 ºC with aeration or at 190 ºC in a fryer. Data taken from [15-17, 21-26]. The signal letter agrees with those in Figures 3 and 4

Signal	Chemical shift (ppm)	Multi-plicity	Functional group	Compound
a	6.58	dddd	-C$\underline{H}$=C$\underline{H}$-C$\underline{H}$=C$\underline{H}$-	(*Z,E*)-conjugated double bonds associated with hydroperoxides
a	6.00	ddtd		
	5.56	ddm		
	5.51	dtm		
b	6.27	ddm	-C$\underline{H}$=C$\underline{H}$-C$\underline{H}$=C$\underline{H}$-	(*E,E*)-conjugated double bonds associated with hydroperoxides
b	6.06	ddtd		
b	5.76	dtm		
	5.47	ddm		
c	8.3-8.9	--	-OO$\underline{H}$	hydroperoxide group
d	9.49	d	$\underline{H}$OC-	(*E*)-2-alkenals
	6.85	tt	HOC-CH=C$\underline{H}$-	
	6.11	dd	HOC-C$\underline{H}$=CH-	
	2.32	q	HOC-CH=CH-C$\underline{H}_2$-	
	1.69-1.19	m	HOC-CH=CH-CH_2-(C$\underline{H}_2$)$_n$-	
	0.89	t	HOC-CH=CH-CH_2-$(CH_2)_n$-C$\underline{H}_3$	
e	9.53	d	$\underline{H}$OC-	(*E,E*)-2,4-alkadienals
	7.09	m	HOC-CH=C$\underline{H}$-	
	6.30	m	HOC-CH=CH-C$\underline{H}$=C$\underline{H}$-	
	6.08	dd	HOC-C$\underline{H}$=CH-	
	2.22	m	HOC-CH=CH-CH=CH-C$\underline{H}_2$-	
	1.47-1.30	m	HOC-CH=CH-CH=CH-CH_2-(C$\underline{H}_2$)$_n$-	
	0.90	t	HOC-CH=CH-CH=CH-CH_2-$(CH_2)_n$-C$\underline{H}_3$	
f	9.54	d	$\underline{H}$OC-	4,5-epoxy-2-alkenals
	6.56	dd	HOC-CH=CH-	
	6.40	dd	HOC-C$\underline{H}$=CH-	
	3.33	dd	HOC-CH=CH-C$\underline{H}$O	
	2.96	td	HOC-CH=CH-CHOC$\underline{H}$-	
	1.65-1.33	m	HOC-CH=CH-CHOCH-(C$\underline{H}_2$)n-	
	0.91	t	HOC-CH=CH-CHOCH-(CH_2)n-C$\underline{H}_3$	
g	9.57	d	$\underline{H}$OC-	4-hydroxy-(*E*)-2-alkenals
	6.84	dd	HOC-CH=C$\underline{H}$-	
	6.33	ddd	HOC-C$\underline{H}$=CH-	
	4.43	m	HOC-CH=CH-C$\underline{H}$OH-	
	1.30-1.70	m	HOC-CH=CH-CHOH-(C$\underline{H}_2$)n-	
	0.90	t	HOC-CH=CH-CHOH-(CH_2)n-C$\underline{H}_3$	

Table 4. (Continued)

Signal	Chemical shift (ppm)	Multi-plicity	Functional group	Compound
h	9.58	d	H̲OC-	4-hydroperoxy-(*E*)-2-alkenals
	8.20	d	HOC-CH=CH-CHOOH̲-	
	6.80	dd	HOC-CH=CH̲-	
	6.33	m	HOC-CH̲=CH-	
	4.66	dd	HOC-CH=CH-CH̲OOH-	
	1.73-1.21	m	HOC-CH=CH-CHOOH-(C$\underline{H}_2$)n-	
	0.89	t	HOC-CH=CH-CHOOH-(CH_2)n-C$\underline{H}_3$	
i	9.60	d	H̲OC-	(*Z,E*)-2,4-alkadienals
	7.09	m	HOC-CH=CH̲-	
	6.30	m	HOC-CH=CH-CH̲=CH̲-	
	6.08	dd	HOC-CH̲=CH-	
	2.22	m	HOC-CH=CH-CH=CH-C$\underline{H}_2$-	
	1.47-1.30	m	HOC-CH=CH-CH=CH-CH_2-(C$\underline{H}_2$)$_n$-	
	0.90	t	HOC-CH=CH-CH=CH-CH_2-(CH_2)$_n$-C$\underline{H}_3$	
j	9.75	t	H̲OC-	n-alkanals
	2.40	dt	HOC-C$\underline{H}_2$-	
	1.61	m	HOC-CH_2-C$\underline{H}_2$-	
	1.32-1.27	m	HOC-CH_2-CH_2-(C$\underline{H}_2$)$_n$-	
	0.88	t	HOC-CH_2-CH_2-(CH_2)$_n$-C$\underline{H}_3$	
k	9.78	t	H̲OC-	4-oxo-alkanals
	2.80-2.67	m	HOC-C$\underline{H}_2$-C$\underline{H}_2$-C(O)-	
	2.47	t	HOC-CH_2-CH_2-C(O)-C$\underline{H}_2$-(CH_2)$_n$-CH_3	
	0.88	t	HOC-CH_2-CH_2-C(O)-(CH_2)$_n$-C$\underline{H}_3$	

d: doublet; t: triplet; m: multiplet; q:quartet.

The concentration of these functional groups increases until it reaches a maximum value; afterwards their sudden degradation is produced, leading to the formation of other compounds, called secondary oxidation compounds. Figure 3 shows that the signals of these functional groups are more clearly defined in the spectra of sunflower oil submitted to a thermo-oxidation process, while showing a shorter average life in virgin linseed oil.

The use of this technique has shown that when the thermo-degradation of these oils is produced under frying conditions (190 ºC, 4 L of oil) in a professional fryer in absence of food, the process evolves in a very different way to the above mentioned. In this case, as Figure 3 shows, neither hydroperoxy nor conjugate dienic functional groups are detected in the spectra of any of the oils studied; it must be noted that these spectra were recorded on a Bruker Avance 400 spectrometer operating at 400 MHz [23-26]. In this case, if these functional groups are formed, their average life is very short and they are not present in the sample in enough concentration to be detected by this technique. Instead of these functional groups, aldehydes are detected from the first minutes of heating (this will be described later in the section devoted to the so called secondary oxidation products). It should be mentioned that studies carried out by other authors at this same temperature, in which the concentration of hydroperoxides was determined by classical methods, such as Peroxide Value, have also confirmed that the concentration of hydroperoxides in oils does not suffer important changes under frying temperatures [59].

The quantification of these intermediate compounds, also named primary oxidation compounds, can be carried out by a simple and easy way either by using standard compounds to obtain absolute concentrations, or by referring their concentration to main components or functional groups present in the sample, such as triglycerides or acyl groups respectively.

It should be added that no other technique allows one to visualize the occurrence of these compounds in the way ^{1}H NMR does. The classical methods are based either on chemical determinations of hydroperoxides through the classically known Peroxide Value or on the absorptions at 233nm or 273 nm in UV spectroscopy; however other compounds having dienic conjugated systems different from the above mentioned can also absorb at these wavelengths.

4.3. Formation and Monitoring of the so Called Secondary Oxidation Compounds

It is considered that in thermo-oxidation processes of food lipids, the primary oxidation compounds generated are intermediate compounds that evolve to form so called secondary oxidation compounds. This latter term refers to compounds of very different natures; nevertheless, only some of them can be observed in the ^{1}H NMR spectrum of the whole sample. Principal among these are aldehydes, whose aldehydic proton signals appear between 9 and 10 ppm, which is a spectral region free of other signals. For this reason, aldehydes can be qualitative and quantitatively characterized by this technique, from the spectra of degraded samples under varied conditions [7, 14-26, 34].

As can be seen in Figures 3 and 4, this technique has shown that, during thermo-degradation of the above mentioned edible oils at 70 ºC with aeration, not only the common n-alkanals, (*E*)-2-alkenals and (*E,E*)-2,4-alkadienals, but also 4-hydroperoxy-(*E*)-2-alkenals, 4-hydroxy-(*E*)-2-alkenals, and 4,5-epoxy-2-alkenals have been detected for the first time [14-17]; all these latter aldehydes, belong to the group of oxygenated *alpha, beta*-unsaturated aldehydes [28].

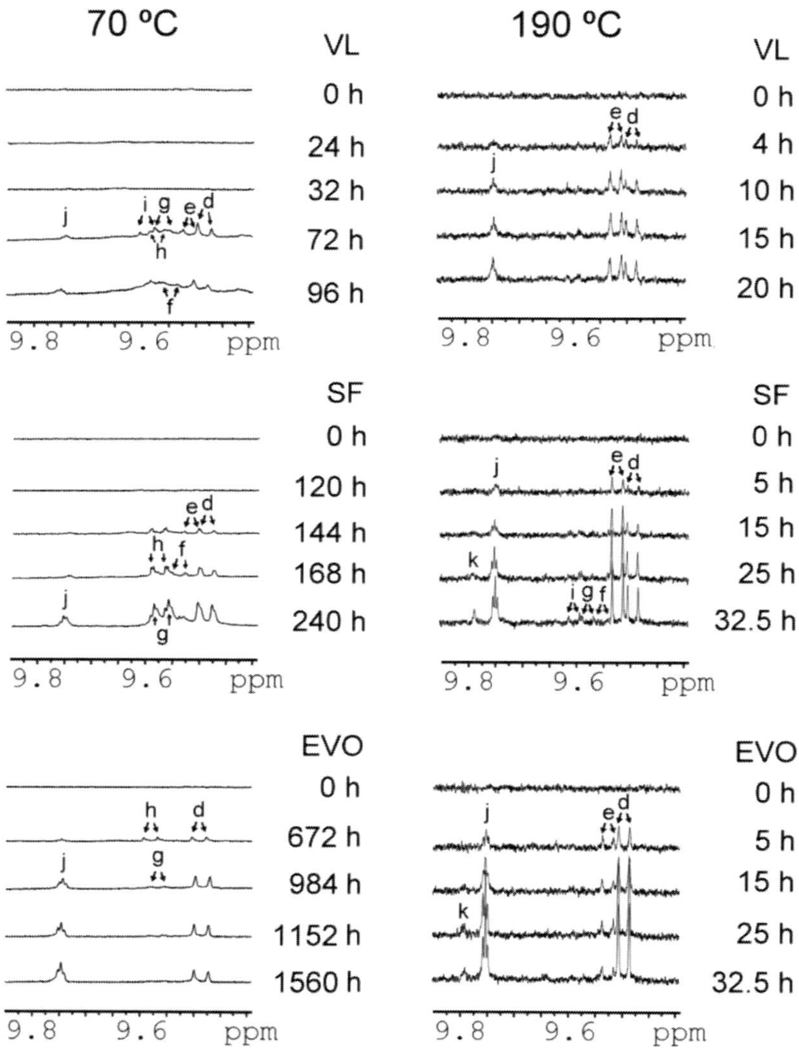

Figure 4. Expanded region between 9.45-9.85 ppm of the ^{1}H NMR spectra (where the signal of aldehydic protons appear) of VL, SF and EVO oils submitted to two thermo-oxidation processes: at 70 ºC in a convection oven and at 190 ºC in a deep-fryer. The signal letters agree with those in Table 4.

The chemical shifts and multiplicities of the signals of these kinds of aldehydes are given in Table 4. These aldehydic groups can be supported on small molecules or can be part of acyl groups in modified triglycerides. In addition, this technique has confirmed that, under the conditions above mentioned, aldehydes are formed mainly during the sudden degradation of primary oxidation compounds. A temporal order in the formation of some aldehydes has been also observed, the concentration of some of them increasing as the others decrease; this is the case of the pairs 4-hydroxy-(*E*)-2-alkenals *versus* 4-hydroperoxy-(*E*)-2-alkenals, and of 4,5-epoxy-2-alkenals *versus* (*E,E*)-2,4-alkadienals.

This technique has also revealed that the first functional groups detectable by ^{1}H NMR spectroscopy during thermo-degradation of the above mentioned oils submitted to frying conditions in the absence of food are aldehydes, and not hydroperoxy or conjugated dienic systems, as Figure 3 and 4 show. As mentioned before, if these intermediate functional groups are formed, their average life under these conditions is very short, in such a way that they are not detected by ^{1}H NMR [23-26, 34]. Under these processing conditions aldehydes of all the above mentioned types, except 4-hydroperoxy-(*E*)-2-alkenals, are also formed, although at different rates and in different concentrations; in addition, another type of aldehyde, 4-oxo-alkanals is also generated under frying conditions. These results show that the conditions under which thermo-degradation occurs can favour or hinder the formation of certain aldehydes.

The quantification of all of these aldehydes can be made either by using standard compounds, obtaining in this way absolute values, or can refer to main components or functional groups such as triglycerides or acyl groups. When these determinations are carried out throughout a process, the kinetic of formation of the several kinds of aldehydes can be studied [23-26, 34]. This technique has proved that under the same degradative conditions not all the oils generate the same kinds of aldehydes, nor is the rate of the aldehydes formation the same in all edible oils. Likewise, it has been shown that the tendency of an oil to generate certain aldehydes depends on the conditions to which the oil is submitted. Among the studied oils, extra virgin olive oil is the oil that generates the lowest concentrations of toxic aldehydes and at the smallest rate, under the same degradative conditions [24-26]. The importance of all these findings lies in the fact that the oxygenated *alpha, beta*-unsaturated aldehydes are geno- and cyto-toxic compounds [27, 28], potential causal agents of degenerative diseases such as cancer, Alzheimer, Parkinson and atherosclerosis among others, and that they could be absorbed from the diet [60-62].

In comparison with other techniques, the advantages that ^{1}H NMR spectroscopy offers in the elucidation of the nature of the aldehydes generated and in their quantification, especially in the thermo-oxidation of edible oils and fats, are enormous in terms of rapidity and simplicity [63]. In fact, by using only this technique the formation of all these toxic compounds under varied degradative conditions has been evidenced.

Other secondary oxidation compounds detectable in the ^{1}H NMR spectra of the whole food lipid sample are epoxides. Their formation has been detected in sunflower oil submitted to different temperatures [22] and also in virgin olive oil [24]. The signal that allows the identification and quantification of these secondary oxidation compounds corresponds to the epoxy protons.

Conclusion

^{1}H NMR is a very useful technique for the study of edible oils and fats, and in general of food lipids, not only from a qualitative but also from a quantitative point of view. As far as characterization studies are concerned, this technique provides a great deal of detailed information about the main components of food lipids, namely triglycerides, and also simultaneously about other minor compounds, like certain minor acyl groups (ω-1), fatty acids, di- and mono-glycerides, phospholipids, terpenes and sterols. Simultaneously, once the signal assignments have been made, the quantification of all these functional groups and compounds can be carried out using different approaches, all of them based on the relation existing between the area of the ^{1}H NMR signals and the number of protons that generate them. Furthermore, this technique is also very helpful to study, from a qualitative and quantitative point of view, the evolution of food lipid degradation processes, providing information not only concerning the evolution of the different kinds of acyl groups, but also about the formation of primary oxidation compounds, such as hydroperoxides and related conjugated dienic systems, and about their secondary oxidation compounds, such as aldehydes and epoxides, some of which have special interest due to their toxicity.

ACKNOWLEDGMENTS

This work has been supported by the Spanish Ministry of Economy and Competitiveness (MINECO, AGL2012-36466), the Basque Government (EJ-GV, GIC10/IT-463-10), and the University of the Basque Country (UPV/EHU, UFI-11/21). A. Martínez-Yusta thanks the EJ-GV for a predoctoral contract.

REFERENCES

[1] Dais, P. & Hatzakis, E. (2013). Quality assessment and authentication of virgin olive oil by NMR spectroscopy: A critical review. *Analytica Chimica Acta, 765*, 1-27.

[2] Guillén, M. D., & Ruiz, A. (2003). Edible oils: Discrimination by ^{1}H nuclear magnetic resonance. *Journal of the Science of Food and Agriculture, 83*(4), 338-346.

[3] Guillén, M. D., & Ruiz, A. (2003). Rapid simultaneous determination by proton NMR of unsaturation and composition of acyl groups in vegetable oils. *European Journal of Lipid Science and Technology, 105*(11), 688-696.

[4] Guillén, M. D., Ruiz, A., Cabo, N., Chirinos, R., & Pascual, G. (2003). Characterization of sacha inchi (*Plukenetia volubilis L.*) oil by FTIR spectroscopy and ^{1}H NMR. Comparison with linseed oil. *Journal of the American Oil Chemists' Society, 80*(8), 755-762.

[5] Schripsema, J. (2008). Comprehensive analysis of polar and apolar constituents of butter and margarine by Nuclear Magnetic Resonance, reflecting quality and production processes. *Journal of Agricultural and Food Chemistry, 56*(8), 2547-2552.

[6] Sopelana, P., Arizabaleta, I., Ibargoitia, M. L., & Guillén, M. D. (2013). Characterization of the lipidic components of margarines by ^{1}H Nuclear Magnetic Resonance. *Food Chemistry, 141*(4), 3357-3364.

[7] Guillén, M. D., & Ruiz, A. (2004). Study of the oxidative stability of salted and unsalted salmon fillets by ^{1}H nuclear magnetic resonance. *Food Chemistry, 86*(2), 297-304.

[8] Guillén, M. D., Carton, I., Goicoechea, E., & Uriarte, P. S. (2008). Characterization of cod liver oil by spectroscopic techniques. New approaches for the determination of compositional parameters, acyl

groups, and cholesterol from ^{1}H nuclear magnetic resonance and Fourier transform infrared spectral data. *Journal of Agricultural and Food Chemistry, 56*(19), 9072-9079.

[9] Igarashi, T., Aursand, M., Hirata, Y., Gribbestad, I. S., Wada, S., & Nonaka, M. (2000). Nondestructive quantitative acid and n-3 fatty acids in fish oils by high-resolution ^{1}H nuclear magnetic resonance spectroscopy. *Journal of the American Oil Chemists' Society, 77*(7), 737-748.

[10] Mannina, L., Sobolev, A. P., Capitani, D., Iaffaldano, N., Rosato, M. P., Ragni, P., et al. (2008). NMR metabolic profiling of organic and aqueous sea bass extract: Implications in the discrimination of wild and cultured sea bass. *Talanta, 77*(1), 433-444.

[11] Siddiqui, N., Sim, J., Silwood, C. J. L., Toms, H., Iles, R. A., & Grootveld, M. (2003). Multicomponent analysis of encapsulated marine oil supplements using high-resolution ^{1}H and ^{13}C NMR techniques. *Journal of Lipid Research, 44*(12), 2406-2427.

[12] Tyl, C. E., Brecker, L., Wagner, K. H. (2008). ^{1}H NMR spectroscopy as tool to follow changes in the fatty acids of fish oils. *European Journal of Lipid Science and Technology, 110*(2), 141-148.

[13] Vidal, N. P., Manzanos, M. J., Goicoechea, E., & Guillén, M. D. (2012). Quality of farmed and wild sea bass lipids studied by ^{1}H NMR: Usefulness of this technique for differentiation on a qualitative and a quantitative basis. *Food Chemistry, 135*(3), 1583-1591.

[14] Guillén, M. D., & Ruiz, A. (2004). Formation of hydroperoxy- and hydroxyalkenals during thermal oxidative degradation of sesame oil monitored by proton NMR. *European Journal of Lipid Science and Technology, 106*(10), 680-687.

[15] Guillén, M. D., & Ruiz, A. (2005). Study by proton nuclear magnetic resonance of the thermal oxidation of oils rich in oleic acyl groups. *Journal of American Oil Chemists' Society, 82*(5), 349-355.

[16] Guillén, M. D., & Ruiz, A. (2005). Oxidation process of oils with high content of linoleic acyl groups and formation of toxic hydroperoxy- and hydroxyalkenals. A study by ^{1}H nuclear magnetic resonance. *Journal of the Science of Food and Agriculture, 85*(14), 2413-2420.

[17] Guillén, M. D., & Ruiz, A. (2005). Monitoring the oxidation of unsaturated oils and formation of oxygenated aldehydes by proton NMR. *European Journal of Lipid Science and Technology, 107*(1), 36-47.

[18] Guillén, M. D., & Ruiz, A. (2006). Study by means of ^{1}H nuclear magnetic resonance of the oxidation process undergone by edible oils of

different natures submitted to microwave action. *Food Chemistry, 96*(4), 665-674.
[19] Guillén, M. D., & Ruiz, A. (2008). Monitoring of heat-induced degradation of edible oils by proton NMR. *European Journal of Lipid Science and Technology*, *110*(1), 52-60.
[20] Guillén, M. D., & Goicoechea, E. (2007). Detection of primary and secondary oxidation products by Fourier transform infrared spectroscopy (FTIR) and ^{1}H nuclear magnetic resonance (NMR) in sunflower oil during storage. *Journal of Agricultural and Food Chemistry, 55*(26), 10729-10736.
[21] Guillén, M. D., & Goicoechea, E. (2009). Oxidation of corn oil at room temperature: Primary and secondary oxidation products and determination of their concentration in the oil liquid matrix from ^{1}H nuclear magnetic resonance data. *Food Chemistry, 116*(1), 183-192.
[22] Goicoechea, E. & Guillén, M. D. (2010). Analysis of Hydroperoxides, Aldehydes and Epoxides by ^{1}H Nuclear Magnetic Resonance in Sunflower Oil Oxidized at 70 and 100 °C. *Journal of Agricultural and Food Chemistry, 58*(10), 6234-6245.
[23] Guillén, M. D., & Uriarte, P. S. (2009). Contribution to Further Understanding of the Evolution of Sunflower Oil Submitted to Frying Temperature in a Domestic Fryer: Study by ^{1}H Nuclear Magnetic Resonance. *Journal of Agricultural and Food Chemistry, 57*(17), 7790-7799.
[24] Guillén, M. D., & Uriarte, P. S. (2012). Study by ^{1}H NMR spectroscopy of the evolution of extra virgin olive oil composition submitted to frying temperature in an industrial fryer for a prolonged period of time. *Food Chemistry, 134*(1), 162-172.
[25] Guillén, M. D., & Uriarte, P. S. (2012). Monitoring by ^{1}H nuclear magnetic resonance of the changes in the composition of virgin linseed oil heated at frying temperature. Comparison with the evolution of other edible oils. *Food Control, 28*(1), 59-68.
[26] Guillén, M. D., & Uriarte, P. S. (2012). Simultaneous control of the evolution of the percentage in weight of polar compounds, iodine value, acyl groups proportions and aldehydes concentrations in sunflower oil submitted to frying temperature in an industrial fryer. *Food Control, 24*(2), 50-56.
[27] Esterbauer, H., Schaur, R. J., & Zollner, H. (1991). Chemistry and biochemistry of 4-hydroxy-2-nonenal, malonaldehyde and related aldehydes. *Free Radical Biology and Medicine, 11*(1), 81-128.

[28] Guillén, M. D., & Goicoechea, E. (2008). Toxic Oxygenated α,β-Unsaturated Aldehydes and their Study in Foods: A Review. *Critical Reviews in Food Science and Nutrition, 48*(2), 119-136.

[29] Gorelik, S., Ligumsky, M., Kohen, R., Kanner, J. (2008). The stomach as a "bioreactor": When red meat meets red wine. *Journal of Agricultural and Food Chemistry, 56*(13), 5002-5007.

[30] Guillén, M. D., & Ruiz, A. (2001). High resolution ^{1}H nuclear magnetic resonance in the study of edible oils and fats. *Trends in Food Science & Technology, 12*(9), 328-338.

[31] Aursand, M., Rainuzzo, J. R., & Grasladen, H. (1993). Quantitative high-resolution ^{13}C and ^{1}H nuclear magnetic resonance of ω-3 fatty acids from white muscle of atlantic salmon (*Salmo Salar*). *Journal of the American Oil Chemists' Society, 70*(10), 971-981.

[32] Sacchi, R., Medina, I., Aubourg, S. P., Addeo, F., & Paolillo, L. (1993). Proton nuclear magnetic resonance rapid and structure specific determination of ω-3 polyunsaturated fatty acids in fish lipids. *Journal of the American Oil Chemists' Society, 70*(3), 225-228.

[33] Miyake, Y., Yokomizo, K., & Matsuzaki, N. (1998). Rapid determination of iodine value by ^{1}H nuclear magnetic resonance spectroscopy. *Journal of the American Oil Chemists' Society, 75*(1), 15-19.

[34] Martínez-Yusta, A., & Guillén, M. D. (2014). Deep-frying food in extra virgin olive oil. A study by ^{1}H Nuclear Magnetic Resonance of the influence of food nature on the evolving composition of the frying medium. *Food Chemistry, 150*, 429-437.

[35] Ando, Y., Ota, T., & Takagi, T. (1989). Japanese sardine oil as a source of 16:3 (n-4) and 16:4 (n-1) fatty acids. *Journal of the American Oil Chemists' Society, 66*(9), 1323-1325.

[36] Skiera, C., Steliopoulos, P., Kuballa, T., Holzgrabe, U., & Diehl, B. (2012). Determination of free fatty acids in edible oils by ^{1}H NMR spectroscopy. *Lipid Technology, 24*(12), 279-281.

[37] Compton, D. L., Vermillion, K. E., & Laszlo, J. A. (2007). Acyl migration kinetics of 2-monoacylglycerols from soybean oil via ^{1}H NMR. *Journal of the American Oil Chemists' Society, 84*(4), 343-348.

[38] Ayer, W. A., & Pedras, M. S. (1987). Metabolites produced by the Scleroderris canker fungus, Gremmeniella abietina. Part 3. Some further metabolites. *Canadian Journal of Chemistry, 65*(4), 754-759.

[39] Zhang, X., Cambrai, A., Miesch, M., Roussi, S., Raul, F., Aoude-Werner, D., & Marchioni, E. (2006). Separation of Δ5- and □7-

phytosterols by adsorption chromatography and semipreparative reversed phase high-performance liquid chromatography for quantitative analysis of phytosterols in foods. *Journal of Agricultural and Food Chemistry, 54*(4), 1196-1202.

[40] Kurata, S., Yamaguchi, K., Ohtsuka, S., & Nagai, M. (2005). Rapid and convenient discrimination of various types of fats and oils by proton nuclear magnetic resonance spectroscopy. *Bunseki Kagaku, 54*(11), 1083-1090.

[41] Mannina, L., Sobolev, A. P., & Segre, A. (2003). Olive oil as seen by NMR and chemometrics. *Spectroscopy Europe, 15*(3), 6,8,10-14.

[42] Perkins, E. G., & Anfinsen, J. R. (1971). Characterization of the nonvolatile compounds formed during the thermal oxidation of triolein. *Journal of the American Oil Chemists' Society, 48*(10), 556-562.

[43] Frankel, E. N., Neff, W. E., & Miyashita, K. (1990). Autoxidation of polyunsaturated triacylglycerols. II. Trilinolenoylglycerol. *Lipids, 25*(1), 40-47.

[44] Neff, W. E., Frankel, E. N., & Miyashita, K. (1990). Autoxidation of polyunsaturated triacylglycerols: I. Trilinoleoylglycerol. *Lipids*, 25(1), 33-39.

[45] Gardner, H. W., & Weisleder, D. (1972). Hydroperoxides from oxidation of linoleic and linolenic acids by soybean lipoxygenase: Proof of the *trans*-11 double bond. *Lipids*, *7*(3), 191-193.

[46] Wineburg, J. P., & Swern, D. (1974). NMR chemical shift reagents in structural determination of lipid derivatives. IV. Methyl *cis*- and *trans*-9, 10-epoxystearate and methyl *erythro*- and *threo*-9, 10-dihydroxystearate. *Journal of the American Oil Chemists' Society, 51*(12), 528-533.

[47] Frankel, E. N., Neff, W. E., Selke, E., & Weisleder, D. (1982). Photosensitized oxidation of methyl linoleate: secondary and volatile thermal decomposition products. *Lipids, 17*(1), 11-18.

[48] Neff, W. E., Frankel, E. N., Selke, E., & Weisleder, D. (1983). Photosensitized oxidation of methyl linoleate monohydroperoxides: hydroperoxy cyclic peroxides, dihydroperoxides, keto esters and volatile thermal decomposition products. *Lipids, 18*(12), 868-876.

[49] Neff, W. E., Frankel, E. N., & Weisleder, D. (1982). Photosensitized oxidation of methyl linolenate. Secondary products. *Lipids, 17*(11), 780-790.

[50] Neff, W. E., Frankel, E. N., & Fujimoto, K. (1988). Autoxidative dimerization of methyl linolenate and its monohydroperoxides,

hydroperoxy epidioxides and dihydroperoxides. *Journal of the American Oil Chemists' Society, 65*(4), 616-623.

[51] Neff, W. E., & Frankel, E. N. (1984). Photosensitized oxidation of methyl linolenate monohydroperoxides: hydroperoxy cyclic peroxides, dihydroperoxides and hydroperoxy bis-cyclic peroxides. *Lipids, 19*(12), 952-957.

[52] Claxson, A. W. D., Hawkes, G. E., Richardson, D. P., Naughton, D. P., Haywood, R. M., Chander, C. L., Atherton, M., Lynch, E. J., & Grootveld, M. C. (1994). Generation of lipid peroxidation products in culinary oils and fats during episodes of thermal stressing: a high field ^{1}H NMR study. *FEBS Letters, 335*(1), 81-90.

[53] Haywood, R. M., Claxson, W. D., Hawkes, G. E., Richardson, D. P., Naughton, D. P., Coumbarides, G., Hawkes, J., Lynch, E. J. & Grootveld, M. C. (1995). Detection of aldehydes and their conjugated hydroperoxydiene precursors in thermally-stressed culinary oils and fats: investigations using high resolution proton NMR spectroscopy. *Free Radical Research, 22*(5), 441-482.

[54] Sheerin, A. N., Silwood, C., Lynch, E., & Grootveld, M. (1997). Production of lipid peroxidation products in culinary oils and fats during episodes of thermal stressing: a high field ^{1}H NMR investigation. *Biochemical Society Transactions 25*(3), 495S.

[55] Silwood, C. J. L., & Grootveld, M. (1999). Application of high-resolution, two-dimensional ^{1}H and ^{13}C nuclear magnetic resonance techniques to the characterization of lipid oxidation products in autoxidized linoleoyl/linolenoylglycerols. *Lipids, 34*(7), 741-756.

[56] Chan, H. W., & Levett, G. (1977). Autoxidation of methyl linoleate. separation and analysis of isomeric mixtures of methyl linoleate hydroperoxides and methyl hydroxylinoleates. *Lipids, 12*(1), 99-104.

[57] Chan, H. W., Matthew, J. A., & Coxon, D. T. (1980). A hydroperoxy-epidioxide from the autoxidation of a hydroperoxide of methyl linolenate. *Journal of the Chemical Society, Chemical Communi-cations, 5,* 235-236.

[58] Neff, W. E., & El-Agaimy, M. (1996). Autoxidation of trioleoylglycerol. *Oleagineux, Corps Gras, Lipides, 3*(1), 71-74.

[59] Dobarganes, M. C., & Pérez-Camino, M. C. (1988). Fatty acid composition: a useful tool for the determination of alteration level in heated fats. *Revue Francaise des Corps Gras, 35*(2), 67-70.

[60] Oarada, M., Miyazawa, T., & Kaneda, T. (1986). Distribution of 14C after oral administration of [U-^{14}C]labeled methyl linoleate

hydroperoxides and their secondary oxidation products in rats. *Lipids, 21*(2), 150-154.

[61] Goicoechea, E., Van Twillert, K., Duits, M., Brandon, E. F. A., Kootstra, P. R., Blokland, M. H., & Guillén, M. D. (2008). Use of an *in vitro* digestion model to study the bioaccessibility of 4-hydroxy-2-nonenal and related aldehydes present in oxidized oils rich in ω-6 acyl groups. *Journal of Agricultural and Food Chemistry, 56*(18), 8475-8483.

[62] Goicoechea, E., Brandon, E. F. A., Blokland, M. H., Guillén, M. D. (2010). Fate in digestion *in vitro* of several food components, including some toxic compounds coming from ω-3 and ω-6 lipids. *Food and Chemical Toxicology*, Volume Date 2011, *49*(1), 115-124.

[63] Martínez-Yusta, A., Goicoechea, E. & Guillén, M. D. (2014). Aldehydes after prolonged heating at frying temperature. In V. Preedy (Ed.), In Processing and Impact on Active Components in Food: A Modern Approach. Academic Press (Elsevier). Oxford, United Kingdom. (In press).

INDEX

C

D

E

F

G

H

I

J

K

L

M

N

O

P

Q

R

S

T

U

V

W

X

Y

Z